The Atmospheric City

The Atmospheric City explores how people make sense of the feelings they experience in and of urban spaces. Based on ethnographic fieldwork of everyday life in Copenhagen, Oslo, and Stockholm, it focuses on the atmospheric power of people, places, and phenomena.

While the predominant focus of current urban planning tends to rest on economic growth, sustainability, or offering housing, transport, and activities to an increasing number of city residents, this book offers a different take, based on recent discussions in the social sciences about how cities *feel*. It calls attention to the mundane ways in which urban dwellers adapt and adopt their surroundings. It argues that cities are characterised by a fundamental atmospheric porosity that affects how people relate to places. This highlights why some places are sought after while others are avoided. Through concrete examples of people being in and moving through the city, the book shows how people attune and are attuned by designed urban spaces, often at the margins of attention, when they find comfort in the familiar and seek out the unexpected.

This book is aimed at researchers, postgraduates, and practitioners interested in urban design and how people make sense of the feelings it evokes. It will be of interest to those in the fields of urban studies, urban design, planning, architecture, urban geography, cultural geography, cultural studies, and anthropology.

Mikkel Bille is Professor of Ethnology at the Saxo Institute, University of Copenhagen.

Siri Schwabe is Assistant Professor in the Department of People and Technology, Roskilde University.

Ambiances, Atmospheres and Sensory Experiences of Spaces
Series Editors:
Rainer Kazig, *CNRS Research Laboratory Ambiances – Architectures – Urbanités, Grenoble, France*
Damien Masson, *Université de Cergy-Pontoise, France*
Paul Simpson, *Plymouth University, UK*

Research on ambiances and atmospheres has grown significantly in recent years in a range of disciplines, including Francophone architecture and urban studies, German research related to philosophy and aesthetics, and a growing range of Anglophone research on affective atmospheres within human geography and sociology.

This series offers a forum for research that engages with questions around ambiances and atmospheres in exploring their significances in understanding social life. Each book in the series advances some combination of theoretical understandings, practical knowledges and methodological approaches. More specifically, a range of key questions which contributions to the series seek to address includes:

- In what ways do ambiances and atmospheres play a part in the unfolding of social life in a variety of settings?
- What kinds of ethical, aesthetic, and political possibilities might be opened up and cultivated through a focus on atmospheres/ambiances?
- How do actors such as planners, architects, managers, commercial interests and public authorities actively engage with ambiances and atmospheres or seek to shape them? How might these ambiances and atmospheres be reshaped towards critical ends?
- What original forms of representations can be found today to (re)present the sensory, the atmospheric, the experiential? What sort of writing, modes of expression, or vocabulary is required? What research methodologies and practices might we employ in engaging with ambiances and atmospheres?

Sensory Transformations
Environments, Technologies, Sensobiographies
Helmi Järviluoma and Lesley Murray

Ambiance, Tourism and the City
Edited by Iñigo Sánchez-Fuarros, Daniel Paiva and Daniel Malet Calvo

For more information about this series, please visit: www.routledge.com/Ambiances-Atmospheres-and-Sensory-Experiences-of-Spaces/book-series/AMB

The Atmospheric City

Mikkel Bille and Siri Schwabe

LONDON AND NEW YORK

Cover image: Muriel de Seze

First published 2023
by Routledge
4 Park Square, Milton Park, Abingdon, Oxon OX14 4RN

and by Routledge
605 Third Avenue, New York, NY 10158

Routledge is an imprint of the Taylor & Francis Group, an informa business

© 2023 Mikkel Bille and Siri Schwabe

The right of Mikkel Bille and Siri Schwabe to be identified as authors of this work has been asserted in accordance with sections 77 and 78 of the Copyright, Designs and Patents Act 1988.

All rights reserved. No part of this book may be reprinted or reproduced or utilised in any form or by any electronic, mechanical, or other means, now known or hereafter invented, including photocopying and recording, or in any information storage or retrieval system, without permission in writing from the publishers.

Trademark notice: Product or corporate names may be trademarks or registered trademarks, and are used only for identification and explanation without intent to infringe.

British Library Cataloguing-in-Publication Data
A catalogue record for this book is available from the British Library

ISBN: 978-1-032-31699-4 (hbk)
ISBN: 978-1-032-45901-1 (pbk)
ISBN: 978-1-003-37918-8 (ebk)

DOI: 10.4324/9781003379188

Typeset in Times New Roman
by SPi Technologies India Pvt Ltd (Straive)

Contents

List of figures vii
Preface viii
Acknowledgements xi

1 Introduction 3

Towards a porous way of thinking 7
The life-oriented city 10
Atmospheric spaces 14
Attuning to atmospheres 16
Exploring the atmospheric city 20
Outline of the book 22
Notes 23

2 Attuned relations: The sociality of atmospheres 25

Chasing resonance 28
Solitude in the atmospheric city 31
Being alone together 35
Dissonant encounters 40
Quiet and loud: Crowds and atmospheres 44
Conclusion 48
Notes 49

3 Embraced by the city: Feeling the urban environment 51

Atmospheric design 55
Feeling the atmospheres of urban design 57
The affective qualities of things 60
Environments of atmospheres past 65
Weathering the atmospheric city 68
Conclusion 71
Notes 72

4 Moving through atmospheres: Mobility and attunement 75

Moved by design 78
Perceiving through movement 81
Moving in practice 85
Moving among others 89
Conclusion 93

5 Cities of care: Nurturing atmospheres 95

Care in the atmospheric city 97
Building care 100
Atmospheres of care 103
Zones of comfort 105
Knowing the city, feeling for the city 108
Caring for plurality 110
Conclusion 114
Note 114

6 The future of the atmospheric city 117

Notes 123

Bibliography 124
Index 137

Figures

1.1	Lights in Gamla Stan, Stockholm. Photo by Siri Schwabe.	2
1.2	Blågårds Plads, Copenhagen. Photo by Mikkel Bille.	4
1.3	Life after dark in North West Park, Copenhagen. Photo: Ida Lerche Klaaborg.	9
1.4	Cold and ice in Copenhagen. Photo by Siri Schwabe.	20
2.1	Coming together at the Oslo Opera House. Photo by Siri Schwabe.	24
2.2	An empty place: Svend Aukens Plads in Copenhagen	26
2.3	A quiet place in the city: Hammarby Kaj in Stockholm	29
2.4	*Amalienborg Plads* by Vilhelm Hammershøi (1896)	31
2.5	The Oslo Opera House foyer	36
2.6	Israels Plads, Copenhagen	38
2.7	Aker Brygge, Oslo	41
3.1	Watching and being watched in Tensta, Stockholm. Photo by Mikkel Bille.	50
3.2	Liquid Light in Oslo	51
3.3	Brunkebergstorg, Stockholm	58
3.4	The oversized lamp in Stovner, Oslo	61
3.5	Illuminated bench in North West Park, Copenhagen. Photo by Siri Schwabe.	63
3.6	Stockholm Royal Seaport	67
4.1	Strolling at the Royal Playhouse, Copenhagen. Photo by Siri Schwabe.	74
4.2	Playing with shadows at the Oslo Opera House	79
4.3	A solitary walker in Stockholm	83
4.4	Cycling in Copenhagen	86
4.5	Sergels Torg, Stockholm	90
5.1	Rooftops in central Copenhagen. Photo by Siri Schwabe.	94
5.2	After dark at North West Park, Copenhagen	101
5.3	Nørrebro, Copenhagen	107
5.4	Being together around the harbour bath at Islands Brygge, Copenhagen	111
6.1	Projected light art at Blågårds Plads, Copenhagen. Photo by Olivia Norma Jørgensen.	116

Preface

Something is happening in the cities of the Nordic countries, but we also see hints of it elsewhere. It is neither a rapid change nor even radically new. But it is seen, heard, smelt, and felt. A change seems to have happened regarding the sensory output of the city from when we were young back in the 1990s. The squares look and feel slightly different. They are no longer 'simply' squares with trees and benches, perhaps a statue, but more like 'theatre stages' with colourful spotlights, artificial hills, neatly planted trees and paths. On the streets, the new lights are whiter and glaring in a different way compared to the light previously illuminating our city in the evening with a yellowish glow. And the sounds and the smells are there, just not in the same overwhelming way as we recall. This is a change that works through *subtraction* and *addition*. There are the sensory subtractions, where smells and sounds are removed: 'Environmental zones', where certain kinds of car traffic are limited, leaving the soundscape more muffled and the air cleaner. The large-scale industries have largely moved out of the city, taking the distinct smells from whatever brewery, slaughterhouse or candy factory that used to cover the neighbourhood, with them. And then there are the additions, in cases where designers and planners design parks, squares, and streets through material and sensuous interventions. Here, urban spaces are, for instance, rendered with colourful spotlights on specific areas, trees, and features in the dark hours. These lights are additions, not simply made to create a *lit* space, but to create a *felt* space. This book is about the city as such felt spaces, and how people make sense, use, and appropriate them in everyday life.

Over the last two decades, a significant portion of urban research has focused on the felt qualities of cities around the world. From tackling the 'experience economy' to situating urban studies within the 'affective turn' and grappling with 'atmosphere studies', a city has increasingly been conceived not so much in terms of what happens in it, but rather in terms of what it feels like. In many ways, this interest harks back to the substantial field of social science and humanities literature dealing with how places are sensed and experienced that emerged in human geography and related fields since the 1970s (Buttimer 1976; Relph 1976; Rodaway 1994; Seamon 1979; Seamon & Mugerauer 1985; Tuan 1974). *The Atmospheric City* is inspired by this plethora of mostly phenomenological studies on urban life, and their approaches have shaped our thinking and that of many others over the last 50 years (for more thorough discussion of place and space see also Cresswell 2004; Malpas 1999; Massey 2005).

At the same time, we find that the book departs from many of these ventures into the urban environment in two ways. Firstly, it narrowly approaches cities as fundamentally atmospheric and highlights how cities unfold as lived spaces by analysing the felt qualities of being and moving together or alone in a world of material and quasi-material phenomena like urban infrastructure, objects, and weather. Secondly, it presents an examination of urban design and development in the Nordic context, which has received increasing, albeit often superficial, empirical attention in recent years. This book should thus not be seen as providing an overview of urban scholarship into the lived and the sensed city, although that would have been a worthwhile endeavour in its own right.

What we set out to do, more modestly, is to add ethnographic depth to this scholarship without losing sight of the importance of design, planning, and political decision-making in shaping cities. With that, we present an atmospheric approach to the city, embracing both its human and its non-human elements, as well as the manners in which these interrelate and overlap. While the focus of much current city planning is often aimed at economic growth, sustainability, or to cater to housing and transport for an increasing number of residents, the abovementioned change is also one that highlights the need to contemplate how cities feel as everyday atmospheric spaces, not least as shaped through design. This is particularly important, when seeing how urban design may shape behaviour and use in often unnoticed and seductive ways.

With that as its starting point, the book shows how people relate to cities by addressing the role of sociality, environments, movement, and care in shaping atmospheric experiences of urban spaces. By exploring this, it highlights how the city affects people emotionally through the porous boundaries between people, elements, and phenomena, challenging a view of the city and its people as clearly distinct entities.

We have aimed in this book to make space for some of the often mundane ways in which people feel their cities and navigate them as *atmospheric* spaces (rather than merely tangible, bounded, and material) in the everyday life of Nordic cities. The reader will thus look in vain for generalising claims or models explaining contemporary urban life, or suggestions as to how urban designers and planners should carry out their work. While empirically offering a regional contextual perspective, we hope to contribute to conceptual and theoretical discussions by painting a fuller picture of urban life as wrapped up in the porosity of human-environmental encounters that make up atmospheres. Simultaneously, we aim for this to offer some much-needed nuance to debates around liveability, not least considering how Nordic cities, in particular, have come to represent widespread notions of the urban good life (cf. Simpson et al. 2018).

In painting this picture, we draw primarily on ethnographic material from our field studies in Stockholm, Oslo, and Copenhagen from 2018 to 2021.

It was carried out by ourselves along with a small group of colleagues at Roskilde University. Rather than making the futile attempt to do justice to the richness of this material, which includes countless hours of participant observation, archival research, and hundreds of interviews, we present selected examples from our fieldwork to think through and illustrate how experiences of urban spaces that may feel highly personal come into being and are collectively shared in the interactions between people, places, and things. While the intention was an equal distribution of examples, it is also evident that COVID-19 shaped our possibilities for fieldwork and thus there is a slight imbalance towards material from Copenhagen. However, we find that the fundamental aspect of how people seek out, avoid, or adapt particular atmospheres runs across the three cities, regardless of this imbalance in representation.

Acknowledgements

This book is based on research generously funded by the Velux Foundation for the project 'Living with Nordic Lighting' [grant #16998]. We want to first and foremost thank Professor David Pinder, Postdoc Anette Stenslund, and PhD student Jeremy Payne-Frank as key members of the project. All three have contributed greatly to the development of the ideas presented here and have provided thought-provoking input to our process over the past four years. In addition, the latter two have graciously offered up their own data and fieldwork-based insights, which we reference and draw upon at various points throughout the chapters of the book. The book, in essence, is thus also a summation and synthesis of the collective work undertaken as part of this project, and some of which has also been published in journal articles and edited volumes by its members. That being said, the book reflects our perspective, and any omissions or misinterpretations remain our own.

We extend thanks to our field assistants, anthropologists Ida Lerche Klaaborg, Olivia Norma Jørgensen, Oda Fagerland, Mikaela Hellstrand, and geographer Johannes Riis, who helped interview the many people who appear on these pages. The first two played particularly active roles in the project and acted as collaborators on publications as well as presentations along the way. Among the many people who have offered their thoughts and support over the years, we are particularly grateful to Jonas Larsen, Sarah Pink, Shanti Sumartojo, Tim Edensor, Tim Ingold, and Tim Flohr Sørensen, whose encouragement, readings, and input have been invaluable. We also thank our colleagues at Roskilde University for their support and feedback, especially the members of 'Lived City', a subgroup of the larger Mobility, Space, Place, and Urban Studies (MOSPUS) research group. Discussions between Mikkel Bille and Mikkel Thelle of urban atmospheres during the COVID-19 pandemic as part of the research project 'The Pandemic City' (also funded by the Velux Foundation) have been particularly fruitful and data from this project is also presented in various sections in this book. Along with Ida Lerche Klaaborg, we thank anthropologists Mie Larsen and Josephine Trojahn for their work as research assistants on that project.

Student assistants Helene Breum, Line Hansen, Jonas Johansen, and Sif Kaad helped transcribe the material. In addition, Carsten P. Sørensen offered his kind administrative support throughout the project period, for which we are thankful.

Copenhagen, August 2022

Figure 1.1

1 Introduction

Mia loved her neighbourhood of Nørrebro in Copenhagen, Denmark. She loved the vibrancy and feeling of diversity that seemed to spill over from all the different restaurants, shops, and people. As a young woman in her late 20s, she knew about the neighbourhood's negative reputation, bolstered by media stories of high unemployment rates, gang-related crime, and occasional riots. And while no doubt this was also part of the story of the neighbourhood to her, she loved it nonetheless. Mia lived alone in a small apartment on a popular square, not far from where she was raised. Being so familiar with the area, she rarely felt unsafe, but when occasionally she did, she would recognise the potential danger and simply choose a different route to avoid it. At the same time, she was often confronted with the concerns of her family in the more rural part of Denmark regarding the neighbourhood. To illustrate the stark difference between the feeling she had of the place and the stories she often heard of it, she recounted a visit from her late grandmother:

> I then took my grandmother down to the square a couple of years ago – she has since died – and at that time she suffered from heavy dementia and couldn't remember all of the stories from the media and so on. I pushed her in her wheelchair on a sunny day full of life in the square and the first thing she said was: 'God, what a wonderful place you live in'.

It *felt* wonderful. Certainly, stories may cling on to a place, regardless of how many design interventions to increase safety have been introduced by local municipalities or others, and indeed the negative stories were an ingrained part of this particular area. But they tend to say little about the experience of actually being there. As we saw above, the dementia she suffered from caused Mia's grandmother to forget the negative stories and simply engage with the atmospheric qualities of the square as it felt on a sunny day.

At the same time, the negative perceptions have been extensively countered in recent years. Copenhagen as a whole was named the world's 'best in travel' by Lonely Planet in 2019. 'Copenhagen is the epitome of Scandi cool. Modernist lamps light New Nordic tables, bridges buzz with cycling commuters and eye-candy locals dive into pristine waterways', they write.[1] From design and cuisine to cycling and sustainability, Copenhagen is 'a city of

DOI: 10.4324/9781003379188-1

endless visual pleasure; a place where even the most mundane activities are laced with a sense of quiet wonder and delight'. The year after, in 2020, the international magazine *Time Out*, covering the ins and outs of city life, ranked Mia's Nørrebro neighbourhood the 28th coolest neighbourhood in the world for its 'never-sleeps sort of atmosphere', where visitors, after a night out, can enjoy 'the perfect morning-after reflection' in the cemetery where fairy-tale author Hans Christian Andersen is buried. In 2021, the neighbourhood went on to top the list, raising obvious questions about the validity of such rankings. In these versions of Copenhagen, it is not simply a city to be judged by things to do or see; about having fun in the second-oldest amusement park in the world, Tivoli, drinking beer by the harbourfront at Nyhavn, eating at world-class restaurants like Noma, or walking on the high street, buying minimalistic Nordic design. Aside from these activities, Copenhagen is essentially about its feeling – *the atmosphere*. It is a place where behaviour and activities are seemingly marked by the relationships between people, urban design, and environmental elements that are infused with atmospheric qualities.

Figure 1.2 Blågårds Plads, Copenhagen.
Photo by Mikkel Bille.

Whether we start from mundane experiences or influential narratives, it is a fundamental condition of human life that people are located in places that have an affective impact, even if their felt qualities are most often perceived at the margins of sensory attention. Places are smelled, heard, tasted, touched, and moved through, imbuing people with affective imprints along the way. There is a wide range of terms and expressions for such sensations of places, from 'energy' or 'vibe' to the notion of there being something 'in the air'. Meanwhile, words often fail to fully capture atmospheres, even if what they cannot describe is somewhat recognisable. Furthermore, while perhaps tempting, efforts to name particular atmospheres may actually risk fixating what is fundamentally ephemeral and give way to a comparison to that which cannot be compared, such as the eerie atmosphere of a COVID-19-stricken city, compared to a deserted playground or an office space in the early morning. These distinct atmospheres may share features, yet are not the same (see Anderson & Ash, 2015).

One feeling that has received a lot of attention in a recent upsurge of international books and media coverage is the Danish term *hygge* (often translated as cosiness), which ostensibly points to a way to relax and live a "happy life". Promoted as the warm, relaxed, and informal atmosphere of a place, *hygge* even reached the national canon of Denmark in 2016 as one of ten central national characteristics, along with the welfare state and gender equality.[2] From these introductory examples, we see that to experience Copenhagen as a city is to engage with the felt qualities of places. And Copenhagen is not exceptional: Every city or place evokes atmospheres. The atmosphere in a place may be carefully choreographed by hosts or designers with music, light, and objects. It may be improvised, adjusted, or erupt in the spur of the moment. Or it may be the outcome of neglect, abandonment, and deprivation. And in Copenhagen, if the claims of *Lonely Planet* are to be taken at face value, such felt qualities lace the most mundane activities 'with a sense of quiet wonder and delight'.

The examples above highlight our basic observation: *Cities are atmospheric*. That is, not merely the sum of material things and bodies that make up the space, but the affective emergence coming together through people, processes, and things. Bringing attention to the feel of a place – or more generally to the city as atmospheric rather than materially fixed and bounded – may at first seem a banal exercise. Of course, places feel different. And we contend with Ash Amin that

> As geographical entities, cities are hardly discernible places with distinct identities. They have become an endless inhabited sprawl without clear boundaries and they have become sites of extraordinary circulation and translocal connectivity, linked to processes of spatial stretching and interdependence associated with globalisation.
>
> (2006, p. 1011)

Regardless of the complexity of delimiting 'a city' as a concept and empirical phenomenon, this book shows that attending to and qualifying such feelings is important. Seeing the atmospheric qualities of places promoted in cities, at restaurants, museums, and by architects and tourism stakeholders to attract visitors, it is becoming increasingly evident that there is potency in and profit to be made from atmospheres. As Anette Stenslund (2023) shows, leading current stakeholders in architecture mostly rest their decisions on metrics and measurements. How much energy will a certain design require, how many visitors might it attract, how many parking spaces should be included, and what are the long-term costs? They may even claim that 'everything can be measured' and seek to turn all 'soft values' into 'hard facts'. Yet, as Stenslund (2023, p. 148) points out, 'while stakeholders may be convinced that it is only numbers that matter, it is the atmospheres that sell'. Numbers may convince, but atmospheres sell. Or as Philip Kotler has noted, 'the atmosphere of the place is more influential than the product itself in the purchase situation' (1974, p. 48). In an urban setting, a focus on metrics carries the risk of overlooking that being in a city is an atmospheric and embodied experience.

At the same time, efforts to detail and plan atmospheres are becoming more and more dominant in urban design practices, with an increase in the number and use of technologies that are employed to shape atmospheres often guided, or legitimised, by quantitative data (Degen & Rose, 2022; Degen et al., 2017; Stenslund, 2023). This is then our empirical point: Urban design is changing, and so is its bio-social affective impact. We see new smart city infrastructures that gather data and monitor uses of space, as well as new specialised AI that combines such data to make evidence-based improvements to built environments; for instance, by employing biohacking lighting technologies to increase hormone production and, ostensibly, well-being, or by lowering noise through implementing sound-reducing materials, improving air quality through traffic regulations, or creating circadian rhythm lighting to improve sleep and productivity. It is not up to us to claim whether such developments are 'bad' in and of themselves. We claim, however, that we are witnessing a tendency for designers, planners, and media to imbue issues of urban mobility, activity, or functionality with a preoccupation with how cities and other spaces *feel*, to improve what is often referred to as 'liveability' (Blanco, 2018; Kaal, 2011; Simpson et al., 2018, Simpson 2019).

Moreover, as the notion of the 'experience economy' has gained prominence, there is indeed an increased focus on evaluating the *experience* of a service, meal, or shop rather than the substance of what is provided. As Anna Klingman notes: 'For architecture, in the experience economy, the relative success of design lies in the sensation a consumer derives from it – in the enjoyment it offers and the resulting pleasures it evokes' (2007, p. 19). In other words, what we see in contemporary cities in the Nordic region and beyond is a focus on experiences and feelings. Yet, there is a danger here of understanding something exclusively by the experience it offers to the subjective perceiver; that is, understanding the city as merely what it feels like and leaving its more formal properties to one side (Klingman, 2007, p. 317). Addressing the atmospheric aspects of the city allows us to move beyond mere experience and address how people, phenomena, and things attune places and affect how they feel while at the same time challenging the numbers-based approach to design. As Mikkel Bille et al. (2015, p. 6) write,

> Focusing on atmospheres means addressing not simply 'experience', but rather the co-existence of embodied experience and the material environment. In order to get to grips with atmospheres, we have to engage more actively and analytically with architecture, colors, lighting, humidity, sound, odor, the texture of things and their mutual juxtaposition.

We likewise posit here that atmospheres are not just something subjectively felt, but also something shared, held in common, and grounded in the materiality of the city. In a broad sense, from this particular perspective, the city *is* its atmospheres. The question, then, is what *qualities* such atmospheres have. What makes a place feel like it does, when, and to whom? A 'never-sleeps sort of atmosphere' can be found in many places. What makes up the qualities of *this* place and what do people feel in it? It is often near impossible to precisely pinpoint and identify an atmosphere, yet we most often nonetheless feel 'something' (Anderson & Ash, 2015). Marketing efforts to describe the felt qualities of a place may rarely match what people experience in reality, simply because people are different and visit places at different times, when different 'energies' dominate. Similarly, people are *positioned* somewhere, or *move* as they experience places, and bring with them past sensory experiences and culturally specific expectations and assumptions that new experiences are set up against. This inevitably influences their attunement to the place, as some senses become more dominant at times or are more intensely activated than others. One may thus experience a kind of calm and soundless urban life if located inside a building looking out at the cars, bikes, and people moving around but without direct access to the soundscape usually expected to go with such a sight (see also Schafer, 1985). Or one may not be accustomed to soundproof windows and come to uneasily ponder the experience of seeing cars but not hearing them. In other words, cities are felt by people as embodied and emplaced beings with specific stories and lifeworlds that engage with the affective mark of place – a mark we call atmospheres. And, as such, atmospheres are integral parts of the world around us.

Towards a porous way of thinking

This book presents an invitation to consider how people make sense of sensuous experiences of the city while taking into account how current developments in urban design risk fixating ideas of how spaces *should* be felt. In foregrounding the atmospheres that leave their mark on urban life, we also call for a move toward what we call a *life-oriented city* that takes into consideration and caters to lived experience rather than relatively static and/or quantitatively based ideas of what makes for good cities. The media, architects, and urban dwellers often talk about cities in atmospheric terms – rather than in socioeconomic terms, for instance – yet the atmospheres that are referred to are not fixed. Their key characteristic, we argue, is that they are *porous*, seeping in and out of people and places. Atmospheres become attuned through the porosity of people, technologies, designs, weather, narratives, plants, animals, and other phenomena. With that, intentionally trying to determine an atmosphere may carry the risk of over-designing places, as it overshadows the plurality in users and uses as well as the social, environmental, and cultural markers that shape urban spaces. It also opens the way for disappointment among users and decision-makers as the gap between the projected future of glossy visualisations and the actualisation of design, shaped by budget cuts, new demands, and the often varying quality of craftsmanship is revealed over time.

The preoccupation with how spaces feel in the field of urban design has analytical implications in that the sensorial and the felt are becoming curated through ambient technologies, calling for the need to draw attention to and nuance our understandings of the city as atmospheric. With that, then, the contribution we seek to make with this book is academic, but we also hope that it might lead to changing practice. We add new empirically informed insights and theoretical perspectives to existing research on urban life and likewise hope to offer practitioners in the field of urban development ideas, concepts, and alternative spatial qualities to *think through*.

At the heart of the issue, what we are interested in is the occasional discrepancy between what cities *should feel* like and how they *are felt*. All cities are atmospheric, but cities are more than what can be designed with atmospheres in mind by planners and architects and promoted in media and travel literature. The atmospheric city is both made up of formal material qualities and experienced by people, who may seek out small squares or places to be alone, be carried away by architecture, or take care to avoid certain areas. Cities consist of tangible things like architecture, infrastructure, statues, water fountains, and human and animal bodies. Cities also come into being as lived spaces through less tangible phenomena like weather, light, smells, and sounds that are more dynamic than material elements such as benches and pavements, yet equally harbour affective and sensorial qualities that have a concrete impact (Rodaway, 1994; Zardini, 2005). Cities, in that sense, have a type of 'atmospherological power' (Hasse, 2014a, p. 223). The urban environment, made up of material and immaterial phenomena, *attunes* people and spurs them on to walk, linger, or haste through spaces, makes them tighten up corporeally or loosen up and relax, and makes them relate to their surroundings in particular ways – even if they do not notice it. The urban environment is active and not just passively receiving projections of subjective affective judgement.

Planning, politics, and economy are obvious aspects of urban life that play a role in constituting the city. Yet, what we point to by attending more narrowly to the atmospheric city is, as Jürgen Hasse writes, that the 'governance of a city never rests solely on political and economic power elites. The citizens and temporary residents also govern the city by living it' (2014a, p. 225). A group of shouting men on a street corner may determine how a space is felt by others much more than any designer could ever determine its atmosphere through arranging lights, benches, pine trees, or separating cars from pedestrians. In that sense, of course, people are more than simply recipients of an affective force, determined by designers at the whim of developers, politicians, and planners. People are not merely attuned *by* atmospheric cities. They also enter spaces as *already attuned* and engage through practices with spaces that in turn change these spaces (Bille & Simonsen, 2021; Edensor & Sumartojo, 2015). Their attunement – the affective mode through which they are present as embodied subjects – may have an impact on how other people around them feel, indeed on how a space as such feels. The observation that the city is atmospheric thus forces us to understand urban space in ways that necessarily take into account both the ability of architecture, infrastructure, time, and weather to affect people, and also how people take part in shaping and evaluating how cities feel depending on their circumstances and biographies. Of course, how atmospheres unfold in Nordic welfare states with particular planning and design traditions may be markedly different in comparison to cities in the Global South or elsewhere. At the same time, although research on atmospheres in non-Western contexts is more scarce (cf. Bille, 2017; Daniels, 2015; Degen et al., 2017; Jaffe et al., 2020; Marinucci, 2019; Tripathy & McFarlane, 2022), it is clear that cities elsewhere are no less atmospheric than in the Nordic countries.

To understand how spaces are imbued with an 'atmospherological power' in the context of the emerging tendency to curate atmospheric aspects of contemporary cities, we delve deep into a range of atmospheric encounters in Copenhagen, Stockholm, and Oslo, while engaging a still-emerging field of research that has focused on atmosphere, affect, and attunement. This field of research has marked a shift in architectural and wider urban theory from a preoccupation with what the built environment 'does' to a preoccupation with what it *feels* like, not simply as a subjective experience, but as something designed and collectively felt. The central questions that guide us are then: *How* do cities feel, to whom, when, and with what implications to everyday urban life?

Figure 1.3 Life after dark in North West Park, Copenhagen.
Photo: Ida Lerche Klaaborg.

The life-oriented city

As a book essentially preoccupied with how cities feel, this is not meant to present an exclusive or holistic approach to the city. The city may be described in sociological or human geographical terminology through a focus on social diversity, increasing gentrification, or the emergence of sociological types, like hipsters, mods, skaters, or joggers. Anthropologists may dive in and describe the everyday life of a specific neighbourhood or street, and historians may bring small stories to life or paint a bigger picture of urban development. Architects may engage with architectural and planning development, imagining new ways of living, while economists and engineers may see the future possibilities of smart cities and look at growth when exploring urban areas. All these perspectives are valid. Yet, all focus on *some* rather than *other* issues, each of which may be relevant to understanding cities. Conversely, we take an interest in life *as such* in the city. To us, exploring urban life is then not just about recognising the presence of people and things, or that there is an 'energy', but also about seeing how life unfolds in an urban context with places for silence, being alone, becoming enlivened, mending heartbreaks, thinking freely, commuting, or feeling at ease. This involves inspiration from psychogeography and the situationist movement, with its explorative approach to looking at the nature of *being in* and moving *through* the city (Coverley, 2018; Knabb, 1995).

Studies concerned with what people *do* in cities have influenced much thinking among urban scholars and practitioners as well as informed design that allows for 'life between buildings' (Gehl, 2011). Starting from the second half of the twentieth century, architect Jan Gehl and psychologist Ingrid Gehl, along with other leading figures, in critiquing the modernist city (cf. Jacobs, 1961; Whyte, 1980; Zukin, 2010), have systematically shown how and where people move and dwell in the city. They highlight through empirical studies how design may help facilitate life in the city, beyond whatever concern or current fad has attracted the attention of planners, designers, and decision-makers, from the 'bicycle city' to 'smart city' over 'sustainability' and 'liveability'. They often focus on attracting people to a place and take this to be a sign of successful design.

Gehl's studies of this sort of 'life between buildings' feature a keen eye for the human scale and perspectives on the city as a living organism, and we consider these influential and valuable studies to be representative of approaches that are predominantly *activity-oriented*: 'Life between buildings is not merely pedestrian traffic or recreational or social activities. Life between buildings comprises the entire spectrum of activities, which combine to make communal spaces in cities and residential areas meaningful and attractive', as Gehl himself writes (2011, p. 14). These ideas are increasingly included into the design of urban spaces on the basis of their offering human perspectives on what is often referred to as *programming*: The implementation of designs based on the notion that cities can be developed in specific ways to better serve people's needs, in particular when it comes to social interaction. As

Gehl suggests, 'public space must be alive, with many people using it' and, continuing: 'The lively city sends friendly and welcoming signals with the promise of social interaction. The presence of other people in itself signals which places are worthwhile' (2010, p. 63). In such a view, a city needs safe spaces for cyclists, pedestrians, and people using alternative forms of mobility; it needs to have benches to sit on and be accessible to both garbage collectors and wheel-chair residents, and all of this needs to make sense in terms of where people are and where they move. Similarly, we see types of activating design that encourage people to *do* something: A bus shelter that also acts as a place to do pull-ups or a park that is not 'just' a park, but also a playground, concert stage, training court, and centre for social and commercial life.

Obviously, activities are of central importance to city life, and the studies mentioned above have provided central contributions to ongoing advocacy for a city designed for people. Yet, they predominantly look at what people *do* in cities and how cities might better facilitate certain activities, in particular social interaction, rather than investigating what these cities *feel* like (and to the extent that they do, it is mostly in terms of 'safety'). Focusing on activities often tells us little about the calmness of a morning stroll, the serenity of sitting on a bench reflecting on the previous night out, or the 'quiet wonder and delight' of such activities; the things that seem to matter not just to international media, but also to people being in, moving through, or moving to the city. In other words, to the feeling of and motivation behind whatever people do in cities. An activity-oriented approach to the city looks at the accomplishment of tasks: Where and when do certain groups of people, most often determined by age or gender, walk, sit, or ride their bikes in a space? It does not necessarily disregard sociocultural factors of feelings in that regard – and Gehl (2011) in particular also works with 12 criteria for a 'good' space, including feelings like safety. Yet, it still relies to a great extent on measurable parameters rather than a qualified attention to the everyday experience of atmospheres.

Complementing this perspective, our claim here is that, from an atmospheric perspective, it is not the *activity* itself that tells us what a lived city is like, but the *attunement* that comes with it. That is, we approach life as unfolding through the atmospheric qualities of a place and how people engage with them, looking not just at what people do, their social interaction, or the density of the local population – although these circumstances may naturally play a part in creating the feel of the place – but at how cities and their spaces feel and come to matter to people. In other words, from the perspective that guides this book, when people go for a stroll, the activity of walking *as such* is not so much in focus. The feeling of fresh air, the sense of city life, the satisfying feeling of tiredness in the body, or the clarification that emerges from thinking freely while moving is what we are interested in. As Gehl (2011, p. 117) himself notes, people who work from home tend to go shopping more often, since the experience rather than the result is the point. It is an invitation not only for shopping but also for contact and stimulation. On the other hand, of course, transportation from A to B is also the point of much activity in the city. Nonetheless, even such purposeful, instrumental practices have an affective aspect to them, whereby the train, bus, or bicycle ride undertaken to get somewhere, feels a particular way, which then becomes part of an overall urban experience. For instance, there is the sense of regularly travelling on the same bus and meeting the same people again and again, knowing that the next morning along the commuter route the bus will again exude the familiar sense of *intimate alienation*. This is the term James Fujii (1999, p. 107) uses for the kind of relationship 'born of repeated sightings in close quarters, one that does not involve or lead to conversation, friendship, or even visible signs of mutual recognition and acknowledgement'. To better understand this and other everyday experiences *in* and *of* urban spaces, we need to focus on how *living* cities feel as they are atmospherically constituted by relations between affective spaces, attuned bodies, time, weather, and other material phenomena.

Approaching the city through atmosphere, in other words, demands a shift in attention. Think not of what people *do* in the city, but of how it *feels* and is *designed to feel*, including through the use of technologies. People we have talked to seek out certain atmospheric qualities in their surroundings through choosing particular routes or places to go and avoiding others. Doing so, they may experience feeling 'mirrored' by their surroundings and feeling at home, comfortable, and safe. This feeling, in turn, is often what makes for a 'good' atmosphere for them. How cities feel cannot simply be traced back to the activities that people perform, or even necessarily to the specificities of what is heard, touched, tasted, smelled, or seen, but is wrapped up in multisensory encounters between attuned beings and their surroundings. This implies a basic shift in ways of thinking about cities toward conceiving them as felt, living entities overflowing with atmospheres (cf. Albertsen, 2019; Bille et al., 2015; Catucci & De Matteis, 2021; De Matteis, 2020; Gandy, 2017; Hasse, 2012; Laage, 2005; Latham & McCormack, 2017; Löfgren, 2015). This book, in that sense, fundamentally presents embodied first-person perspectives on how cities appear to people through atmospheres, and how they make sense of them: How atmospheres are a basic urban condition that has a bearing on their way of life.

Some readers might say that there is nothing new in this attention to people, and indeed humanistic geography (Buttimer, 1976; Relph, 1976; Seamon, 1979; Seamon & Mugerauer, 1985; Tuan, 1974) and psychogeography (Coverley, 2018; Ellard, 2015) have yielded many similar insights. Similarly, ideas of 'designing for people' and concepts such as 'human-centred' or 'user-centred' design have dominated for some time (cf. Giacomin, 2014; Madsbjerg, 2017; Madsbjerg & Rasmussen, 2014). Undoubtedly these approaches have added much-needed insights into what it means to live with things and technologies, most often because they engage the everyday life of the participants. Implemented into design and technology practices, however, similar approaches rest on a particular understanding of evidence that renders the question of what it means to be human quantifiable, offering metrics of effects on a human *body* of, for instance, 'improved' light, sound, air, etc.

To illustrate this, in lighting design, the notion of 'human-centric' lighting currently implemented across public and private architecture and spaces, is essentially about mimicking daylight in order to impact physical bodies. It is about causing an effect that improves the biological aspects of human life. When at its best, 'human-centric' lighting is about delivering 'a specified set of visual, biological and behavioural responses identified as appropriate for the users of that lighting' but: 'At its worst, human-centric lighting is a hollow phrase used to healthwash lighting products or lighting solutions' (Houser et al., 2021, p. 98). In such quantified versions of human-centric design, 'being human' is being reduced to a passive body that design or technology can act upon to make it more productive, healthy, or efficient. Several scholars have critiqued this approach to the 'human-as-body' for its quantitative understanding of being human, but also for traditionally taking the human being to be an able-bodied man and, in extension, critiqued urban environments for catering primarily to this figure (Eichler, 1995; Kern, 2021; Rendell et al., 2000).[3]

In that sense, while not necessarily without any merit *per se*, we believe that a human-centric approach in its quantified form presents a limited and limiting view of human life. In contrast to this quantified version, another perspective on the human is found in ethnographic approaches where a human is a person with social relations, placed in particular spatial and political contexts (Madsbjerg, 2017; Madsbjerg & Rasmussen, 2014). In extension of this, we are preoccupied with a kind of 'human' that is entangled in atmospheric worlds in ways that may range widely from including feelings of attachment, immersion, and being carried away by the 'energy' in a space, to detachment, alienation, and suspicion (see also Stewart, 2011).

This ability of the environment (both built, designed, and natural) and humans to seep affectively in and out of each other, is central to the atmospheric city. Analytically, we call this *the porosity of the atmospheric city*. The porosity of both the city and human life has been particularly evident during the COVID-19 pandemic. Urban dwellers have become increasingly aware that traces of people are left in the air, on benches, and door knobs and handles, only to linger for hours after their passing by and intermingling with other people's bodies as they interact with these same materials. In essence, porosity is a central feature of a *life-oriented city*. The 'city' in this perspective is seen as lived and as alive, composed of relations between affectively and biologically porous beings and entities (see also Stavrides, 2018). During COVID-19, much attention was paid to this porosity as a potentially contagious relation, as the virus moved between bodies. But we argue that this porous relationship extends far beyond this limited period and entangles emotional and biological processes with architecture, design, and matter in a continuous process. In that sense, the porosity we are concerned with is a characteristic of urban life that includes yet also goes beyond the architectural setting and the spatially contained geographic location (see Wolfrum et al., 2018). Porosity, we argue, is thus a core characteristic and defining factor in shaping the atmospheres through which cities are felt.

Atmospheric spaces

Atmospheres, as illustrated above, play a central role in the experience of cities but are rarely as polished and neatly physically bounded as they tend to be presented via CGI renderings by urban designers or photographs in the media. In Nordic cities and others, it often rains, there are smells, and city life may at times be considered a stressful and conflictual experience (Brighenti & Pavoni, 2019; Corbin, 2014). In a basic sense, we see atmospheres as the felt qualities of spaces that arise in the meeting between humans and the environment. These qualities are often intentionally staged through designed objects, lighting, sounds, and the like, not unlike what we know from theatres and cinematography (cf. Groening, 2014; Spadoni, 2020). Yet it is when people are present that places take their shape as experiential spheres: *That* there is an atmosphere in any given place and *what* this atmosphere is are two different questions, and the latter very much hinges on human input. Or, as Dreyfus summarises (albeit by using the term 'mood'): 'The mood *in* a room depends on a group of people sharing the mood, while the mood *of* a room, say, tranquil, energising or oppressive, is there even if no one is in the room' (2012, p. 33). In a theatre, the atmosphere is shaped by the stage setting, the actors in relation to the script, and the kinds of guests present, including how they look, their behaviour, and the expectations that these guests may have based on reviews, promotional material, or otherwise. At the cinema, a film may be visually and narratively processed to feel a certain way, but the film experience rests on whether guests are completely immersed in the plot and not noticing the people around them, are laughing *with* others without really knowing why, or whether the immersion is broken by the annoying disturbance from popcorn, mobile screens, and talkative people (Hanich, 2021). The atmosphere in the theatre and cinema is thus collectively felt and shaped, although the individual attunement to such atmosphere is subjective.

In an urban space, the atmosphere becomes even more difficult to stage, as there are most often multiple factors involved in its becoming: Municipality rules, private-public boundaries, commercial interests, traffic, weather, seasons, times of day, levels of maintenance, not to mention the impulsivity of people (and animals) to act in certain ways. Furthermore, urban space, as lived space, is not straightforwardly defined by material borders but is likewise marked by temporal rhythms and ephemeral layers of felt qualities (cf. Degen & Lewis, 2019; Hasse, 2014b, 2018; Löfgren, 2014). In a sense, you can walk into a pub and detect a joyous atmosphere in one corner around the band, whereas you might find a calmer atmosphere of concentration, competition, and more relaxed sound levels in the corner furthest away where the darts players gather. A physically demarcated space may thus be momentarily occupied by qualitatively different atmospheres. One question is then whether we have *one* atmosphere with 'sub-atmospheres' or simply altogether *different* atmospheres in that space. We will in this book refrain from taking a general stance on this philosophical issue and simply offer perspectives on it from specific ethnographic contexts. As we begin to explore the porosity of atmospheres – of how affective qualities seep in and out of bodies and things – borders and demarcations start to blur in any case, and the question somehow seems to lose some of its initial traction and empirical grounding.

The theoretical implication of taking seriously the felt qualities of a space, according to the phenomenological tradition, is that you do not feel – in the singular – the surface of the drain grates, the tree trunk, the bicycle rack, or each of the moving people around you. You feel the affective impact of a vibrant square made up of an assemblage of various phenomena, things, people, and animals, all presenting you with textures and affective qualities of 'tempered' or 'tuned' spaces (Binswanger, 1933, 618 pp.; Bollnow, 1963, p. 230). In that sense, we might, with Gernot Böhme (1995, 2017), take atmospheres to be that distinct albeit hard-to-grasp spatial aspect of experience that lends both depth and immediacy to people's encounters with the world (see also Brennan, 2004). As Böhme notes,

> Atmospheres are indeterminate above all as regards their ontological status. We are not sure whether we should attribute them to the objects or environments from which they proceed or to the subjects who experience them. We are also unsure where they are. They seem to fill the space with a certain tone or feeling like a haze.
>
> (1993, p. 114)

Following from that, atmospheres do not merely belong to the realm of individual experience but form part of collective situations, even if they can be felt as intensely personal (Anderson, 2009, p. 80). In this sense, 'atmospheres are affective phenomena, which are grasped pre-reflectively, manifest spatially, felt corporeally, and conceived as semi-autonomous and indeterminate entities' (Trigg, 2022, p. 3). The implication of this is that they are 'there but not there, imperceptible yet all-determining' (Philippopoulos-Mihalopoulos, 2013, p. 36). Summing up, Ben Anderson notes that they are a type of experience 'that occur *before* and *alongside* the formation of subjectivity, *across* human and non-human materialities, and *in-between* subject/object distinctions' (2009, p. 78).

Of course, we may come across an atmosphere that is strongly tied to a place, a sort of *genius loci*, a spirit of the place (Norberg-Schulz, 1980) that helps us understand how one place is markedly different from another. As Dylan Trigg (2022, p. 4) suggests,

> The atmosphere is not 'activated' upon our arrival, but instead belongs to the place and inheres in the materiality of the environment. Thus, when we leave that city, place, or room, the atmosphere does not leave with us. Rather, it remains embedded in the things of the world, such that other people can attune themselves to that atmosphere long after we have left.

However, even in places where atmospheres are so pervasive that they seem almost inherent to them, all atmospheres, following Böhme's (1995) approach, emerge in the interplay between subjective bodies and the world of objects and phenomena. There may be a sense of monumentalism and grandeur in the museum districts of Vienna, Berlin, or Paris. Yet, rain, traffic, riots, or particular odours may change the atmosphere instantly. Atmospheres are thus felt in place, yet are also changeable. They are part of an experience, materially constituted, but they are also a conceptual tool that helps us understand that which may not be verbalised or even drawn attention to, but which still shapes people's everyday affective lives.

To capture this complexity of atmosphere as both an experience and an analytical concept, Shanti Sumartojo and Sarah Pink (2019) make a point of distinguishing between knowing *in*, *through*, and *with* atmospheres. As researchers we interact with our participants *in* atmospheres, trying to learn about their world-view and experiences. This raises methodological questions 'as we come to make sense of how other people experience the world, how these experiences come to have meaning and what these meanings are' (2019, p. 11, see also Thibaud, 2013). Researchers like ourselves also think *about* atmospheres, trying to describe them and interpreting their meaning *after* the experience of our participants. And finally, researchers think *through* atmospheres to understand other aspects of human lives, seeing how atmospheres are entangled in ideas about 'the good life', morality, politics, and sociality, just to name a few. In that sense, attending to atmospheres offers both a way of being in place and a way of knowing a place.

Attuning to atmospheres

Much of the work on atmospheres mentioned above builds on Martin Heidegger's related notion of *Stimmung*. In the phenomenological tradition, *Stimmung* is a fundamental mode of existence constitutive of the way one finds oneself in the world (*Befindlichkeit*).[4] It is variously translated into 'attunement' or 'mood', yet should be understood as something radically different from an internal, mental state, referring rather to a condition of being in the world. People are 'always already attuned', as he notes (Heidegger, 1996, §29). We will use the term 'attunement' for *Stimmung* (along with the related verb 'attune') and follow Heidegger as he posits that attunement

> is not some being that appears in the soul as an experience, but the way of our being there with one another [...] Attunements are the fundamental ways in which we find ourselves disposed in such and such a way.
>
> (1995, pp. 66, 67)

He continues,

> Attunements [*Stimmungen*] are *not side-effects*, but are something which in advance determine our being with one another. It seems as though an attunement is in each case already there, so to speak, like an atmosphere

in which we first immerse ourselves in each case and which then attunes us through and through.

(1995, p. 67)

As will be discussed in more detail in Chapter 2, according to Heidegger this fundamental attunement, where the distinction between self and world is dissolved, has the implication that people engage and find themselves *being-with* their environment rather than simply being *in* it as some external entity, and that people are thus also always in the process of being attuned *with* the world (Hasse, 2019). Put perhaps a bit more simply, the world that surrounds people and their mode of being in it cannot be separated; both take shape in a process of mutual response from this phenomenological perspective. 'Basically, there is no condition of human life that is not already attuned in a certain way', as Otto Bollnow notes (1941, p. 54, our translation), whether we are aware of it or not. Someone may indeed be highly aware of their lifted energy on a spring day or completely unaware of their hectically moving, contracted body while they rush to the supermarket for groceries after leaving the office late, even as their disposition shapes their shopping experience.[5] When we talk about attunement, we thus talk about the way people find themselves in the world as already attuned, where separations between emotions, feelings, and affects are blurred in 'an intimate, compositional process of dwelling in spaces that bears, gestures, gestates, worlds' (Stewart, 2011, p. 445).

People's attunement may shift according to the atmospheric qualities of their surroundings, and producing such an effect in people through design, as we touched upon above, may be an active strategy for commercial or experiential reasons (cf. Allen, 2006; Degen & Lewis, 2019; Edensor, 2015). This kind of shift may likewise be caused by a charismatic person, or persons, entering a space and changing the feel of it. Like when your favourite band finally enters the stage to a collective outburst of euphoria following a period of relative boredom and increasing impatience with toe-stomping fellow concertgoers, rude bar staff, and having to stand up and listen to the supporting act with indifference. Or it may happen through movement, for instance, when you find yourself emotionally invigorated by opening the door to a party or when you enter a quiet graveyard that may scare you, calm you down, or bring out a sense of melancholy. In simple terms, we see attunement as the process of engaging with the felt qualities of the world. In that sense, the essence of attunement is the processual entanglement of human bodies and the atmospheric worlds of other people, sentient beings, materials, media, and substances (see also Howes, 2019; Ingold, 2000, 2011).[6]

This observation has an important implication. Being attuned to – and taking part in shaping – environments affords people with affective lenses through which they perceive their surroundings. However, being part of an atmosphere does not entail that people are necessarily swept away by it, or that this atmosphere can ever be somehow fixed. As we have already pointed out, interior designers, architects, or a dinner party host may seek to curate an atmosphere, but they cannot determine the actual, lived outcome of their design. Atmospheres and attunement are thus not the same, even if they are closely related; 'we can behold atmospheres from a distance, sense their presence without being ourselves in their affective grip' (Slaby, 2020). Atmospheres come into being and dissolve, they emerge, consolidate, and coagulate, and people may be differently taken by them in any of these states. Atmospheres are also porous. They seep in and out of people, making affective imprints along the way while simultaneously being affected by the presence of attuned people, things, and phenomena. An atmosphere in a place is thus distinct from someone's attunement to it. As Jan Slaby (2020, p. 275) notes, we 'cannot do without such a distinction, however vague and porous the boundary between atmosphere and corporeal attunement may be in each given case'. Thus, atmospheres are neither located in the subjective mind and body, nor solely outside in the design of things. Rather, they are the relational phenomena that continuously emerge between people's individual attunement and the atmospheric qualities of other people, places, things, and phenomena (Böhme, 2017; Griffero, 2014).

One of the most telling ways in which people attune and become attuned is through feelings of resonance and dissonance. Bodies resonate with one another and with the situation they find themselves in, while 'space is given by the (at times dramatic) interrelation between what affects the subject and the way he responds to it' (De Matteis, 2018, p. 437). Through these relational dynamics, people become attuned to their surroundings and attune them in turn, for better or worse. Indeed, the idea of resonance and dissonance involves a view of the body as *immersed* in the environment (Thibaud, 2015; see also Rosa, 2019). The resonant person is someone who is not separated from their surroundings but who leaves affective traces and is fundamentally open to their surroundings. And vice versa, dissonance erupts as a separation between attunement and atmospheres emerges. This further highlights how porosity is central to our approach to the city. As attuned beings, the boundaries around people's bodies are constantly moving and inherently tenuous.

What is more, how people attune (to) spaces often requires a subtle kind of action, what Sara Ahmed (2014) calls 'mood work'. When something is not quite right, when the experience of resonance is disrupted by the presence of someone or something unable or unwilling to seamlessly become attuned to an existing atmosphere, work is required to close 'the gap between how one does feel and how one should feel' (Ahmed, 2014, p. 21). While this labour (and whether it is successful or unsuccessful) can often have political implications, it is also something that occurs in the everyday ways of attempting to tweak one's bodily presence or disposition to 'fit in' or in ways of setting up spaces to reflect or engender a certain atmosphere, for instance, through arranging lights in a certain way or filling a space with the smell of incense, a scented candle, or baked goods (see Bille, 2019; Rodaway, 1994; Watson et al., 2021).

However, and importantly, there is a fundamental unevenness in regard to who gets to play a role in constituting collective atmospheres and who, on the other hand, is required to do the work needed to tune themselves into them. To illustrate this point, Ahmed takes a description by bell hooks of what happens as a woman of colour joins a room full of white feminists. Ahmed argues that the woman of colour disrupts their pre-established sense of bonding and togetherness by simply bringing herself, her body, her history, which does not align with what is already present, into the room. Not only is there a lack of attunement at her end, she also becomes the cause of tension and, with that, loss of attunement among the others present. As a consequence, it falls on the newly arrived woman to make up for this by labouring to recreate the lost atmosphere (2014, p. 22). She becomes, in Ahmed's terminology, a 'stranger', someone who does not automatically belong, someone who must work to become attuned and is a source of (potential) disruption until that happens. As a corollary to that, one of Ahmed's central points is that, while attunement implies a certain openness to one's surroundings, it does not necessarily imply openness to just anything. Indeed, 'attunement is not exhaustive' (2014, p. 17). When something is 'not in tune' with us, we might block it out, 'close off our bodies' to it (ibid.). Likewise, when someone or something is registered as strange or alien, there may be a loss of attunement and a sense of frustration in the wake of it (cf. Nörenberg, 2018). With that, a refusal to become attuned can also be a political act; to claim, as Ahmed (2014) does, that one is 'not in the mood' – in our vocabulary not attuned – is also to challenge the atmospheres people are expected to work to attune themselves to.

In this focus on attunement to atmospheres, we find the seeds for a radically different way of thinking. What lies beneath the basic question 'have you ever felt moved by a place?' is the counter-intuitive logic that feelings are not merely 'inner' subjective processes in the brain projected *onto* the world, but may also in many cases be external forces that seize the human body and are made sense of (Casey, 2022). In this phenomenological approach, the landscape is not serene because that is the idea that you *project* onto it, but it is the other way around: It is serene because that is how it makes you feel (for more see Böhme, 2017). Similarly, the elevator is claustrophobic because that is how the space grips *your* body, not because you simply project the idea of confinement onto it. This is not to say that feelings are *only* external or externally triggered. Anger, envy, or sorrow can be feelings projected towards an external object, consuming all aspects of a person's presence in a given situation, but social codes may require them to refrain from showing it, and make that feeling more or less invisible to others. On the other hand, a nervous person may be speaking in public with a shivering voice and contracted bodily expression and be felt as intensely cringe-inducing by listeners (cf. Slaby, 2020, p. 278).

Finally, as a mode of tuning-in and tuning-with the world, attunement points to the overall 'relational environment' that people find themselves in at a given time (Manning, 2013, p. 11). This environment, as we will continue to highlight throughout this book, is always marked and perceived in one way or another through atmosphere. It is also made up of presences beyond the human; in the public spaces of cities, people often take particular note of the built environment and its materials. Yet, environments are also made up of numerous other phenomena that leave sensory impressions on people: The weather, the plant and animal life, the noise arising from nearby encounters – friends chatting, music resonating between soundboxes and whatever obstacles get in the way of the soundwaves, skateboards meeting pavement – and the sensory impressions that often lay bare what activities are taking place in their vicinity, from cooking to cleaning to waste management or the lack thereof. The relational and processual character of atmospheres and attunement allows us to identify less programmed ways of being in cities and to observe how the fluidity and ephemerality that comes with the continuously changing environment, not least through the weather, season, and time of day, facilitate a broader spectrum of engagement with space.

Figure 1.4 Cold and ice in Copenhagen.
Photo by Siri Schwabe.

Exploring the atmospheric city

At this point, a note on methods is warranted. Like many others before us, we have faced challenges in approaching the realm of atmospheres as lived phenomena. Being intangible and always changing, atmospheres are by definition difficult (if not impossible) to grasp in any straightforward way; they tend to resist attempts to pin them down, whether in how they are described by those who experience them or how they are known by social scientists like ourselves. Studying atmospheres is perhaps most fruitfully done by taking seriously their slipperiness, their porosity, and the haziness with which they often seem to present themselves to us. Taking inspiration from others who have grappled with the methodological issues that arise from atmospheres (Adey et al., 2013; Anderson & Ash, 2015; Ash & Gallacher, 2015; Edensor & Sumartojo, 2015; Sumartojo & Pink, 2019), and continuously building on our own considerations in that regard, our approach to the atmospheric city has necessarily been both flexible and open-ended, while retaining an ethnographically founded methodological rigour. This approach fits with the overall aim of this book, which is not to pin down urban atmospheres and create a generic picture of a fixed range of atmospheres but rather to allow the complex and elusive nature of atmospheres to guide us toward a better understanding of lived cities.

This of course leads us to another, related predicament: People never experience urban space unmediated, whether as dwellers, visitors, workers, or researchers. James Donald reminds us that '[t]he city we do experience – the city as state of mind – is always already symbolised and metaphorised' (1999, p. 17). In other words, none of us can ever hope to approach the city as a social space truly objectively. To most (social) scientists, this will come as no surprise; for decades, we have continuously and critically reflected on our positions, our preconceived notions, and how these shape our impressions of the world, our thinking processes, and not least the ways we end up representing what we find. For the purposes of this book and the research behind it, considerations on ethnography as an embodied practice have also been central. How can we even begin to think productively of the atmospheric city without acknowledging the subtle attunements as well as experiences of resonance and dissonance that shape our own relationships to our surroundings and to the participants whose lives we briefly interrupted with our research? In the chapters that follow, our own observations (and those of our research team members) have been held up against the way our research participants make sense of spaces – people who we simultaneously share a range of similarities with and whom we are very different from. These similarities and differences, in turn, have prompted us to ask new questions along the way. The chapters have also been thrown into fruitful conversation with the manifold observations, insights, and wider perspectives offered by our colleagues, our research assistants, and the wide array of scholarly literature that has helped guide this project.

The empirical material on which we base our analysis consists mainly of 272 interviews by ourselves and our team members with residents, users, designers, and visitors of urban spaces in Oslo, Stockholm, and Copenhagen that were carried out with a particular focus on atmospheres and lighting from 2018 to 2021 as part of a research project on 'Living with Nordic Lighting'. Our participants ranged in age from 12 to 89 and included 165 women and 119 men, as well a non-binary person and two people whose gender was not identified. The conversations we had with our research participants include in-depth interviews in people's homes and workplaces, interactions during recorded walks, as well as formal and informal interviews and voxpop-style surveys conducted in parks, squares, and other public spaces. In addition, we and our team have engaged in a range of methods, from participant observation and extensive note-taking, tracing and tracking uses of particular spaces using maps and tables, to audio-visual approaches and more experimental exercises with podcasts, short films, and pinhole photography. Importantly, while we have undertaken the task of writing this book, the insights presented spring from the collective effort of a larger team of researchers, who have spent numerous hours becoming involved in people's lives, building trust, and documenting their work throughout the process.

Outline of the book

In the following chapters, we present what we consider the most pressing perspectives on the atmospheric city as we have come to know it in the Nordic context. The relational quality of atmospheres and their porosity run as red threads through the chapters, which are organised according to four central themes: *Social relations*, the *environment*, *movement*, and *care*. These four themes are not in any way exhaustive, of course, but they have been the most dominant themes arising from our material, both as constituent elements of lived, atmospheric cities and as useful conceptual paths into them. Just as they do not form an exclusive list, we do not attempt to paint a generic and universalising picture with them. Rather, we use them to present observations that we believe are both timely and conducive to a richer understanding of urban life as it unfolds in the atmospheric spaces that arise between everyday attunement and the ongoing shaping of urban environments through design and technologies, ultimately pointing toward a life-oriented city.

First, in Chapter 2, we delve into the *social relations*. When you find yourself immersed in an atmosphere, are you alone or are there other people around? Who are these people and what are they doing? The chapter focuses on the social relations that shape atmospheres and attune the city and its people. In other words, we focus on the human interactions – or lack thereof – that feed into and are fed by atmospheres. While design and technology may frame human activities, we show how they cannot entirely determine how places feel or why people go there, since atmospheres, aside from material media, rely on social relations and interactions.

In Chapter 3, we look further into the second theme, the *environment*, both material and non-material, that constitutes an inescapable component of atmospheric cities. Here, we dive into the city as made up of non-human presences, from natural elements to infrastructure and the various objects and phenomena that surround urban dwellers. These, we argue, are central to understanding the atmospheric city in that attunement necessarily takes place through corporeal relations with the non-human, both material and non-material, animate and inanimate.

In Chapter 4, we shift our focus to the role of *movement* in attuning human bodies, and cities with them. We explore how people move in and through urban spaces, and ask how movement influences how they experience atmospheres and make sense of the environments they find themselves in. Different ways of moving imply different rhythms and paces, and facilitate different engagements with atmospheres. Meanwhile, both walking and cycling, as the two dominant ways of movement that we explore in the chapter, may also offer time and opportunity for entering a meditative state, or for feeling a rush while moving quickly with others. These two ways of movement are thus not the same, yet both involve processes of attunement that may be shared across mobility forms.

In Chapter 5, we investigate how ideas and practices of *care* manifest themselves in how cities are felt. This focus on care is a result of our attempts to come to terms with how attunement comes to characterise particular forms of relationality between and beyond humans. Care, we argue, is not simply another mode of engaging the atmospheric city but is perhaps better understood as an overarching quality of that engagement, and whether as caring or cared for, the chapter offers a glimpse into what the future of a more life-oriented way of thinking about the city might hold.

Following from that, we round off with a concluding chapter (Chapter 6) in which we gather the threads from the preceding chapters and reflect on the way new technologies and emphasis on data influence city planning and design.

Notes

1 https://www.lonelyplanet.com/denmark/copenhagen. Accessed 10 June 2020.
2 The full list (in Danish) can be found here: https://www.regeringen.dk/nyheder/2016/danmarkskanon-befolkningen-har-valgt-10-vaerdier-for-fremtidens-samfund/. Accessed 10 June 2020.
3 For further perspectives beyond urban design, see Criado Perez (2020).
4 We will not go into further details here, but simply note that there is a difference between the notion of *Stimmungen* and *Befindlichkeit*, in Heidegger's writing, both often translated into attunement or mood.
5 In that sense, this is a different take on attunement/mood than one may find in medicine or psychology books, where mood is contrasted with 'emotion' under the umbrella term of 'affect'. Here, moods and emotions are distinguished by their durations and whether they are directed *at* something. Moods are in this literature transient, low-intensity, and nonspecific without a cause, while 'Emotions are fairly fleeting and intense experiences that are elicited in response to specific external stimuli (i.e., objects or events)' (Niven, 2013, p. 49).
6 We attempt here to communicate a somewhat clear distinction between much discussed terms. This, however, inevitably leaves behind some nuance that has been intensely discussed in academic circles, and the interested reader may therefore find further philosophical clarification here (cf. Ferran, 2022; Griffero & Tedeschini, 2019; Griffero, 2014, 2021; Hasse, 2019; Slaby and von Scheve, 2019; Trigg, 2022).

Figure 2.1

2 Attuned relations
The sociality of atmospheres

As COVID-19 spread across the world in 2020, it became apparent that the idea of the body as a tangible, buffered container with a so-called 'inner mind' in contrast to an 'outer world' is somewhat imperfect. Indeed, parts of people's biological bodies are left behind on door handles, benches, supermarket carts, and in the air they breathe. And parts of other people filter through these spaces and bodies. A person is, in the words of Charles Taylor, a 'porous self' (2007, p. 38). You may know the feeling of stepping into an empty elevator and being confronted by the overwhelming smell of stale cigarette smoke, perfume, or a lingering fart. You may cast moral judgement upon the now absent person, or even come to characterise the place itself as 'dirty', as you defenselessly take in something of that person through the air. This bodily porosity may also allow for undetected circulation of pheromones interfering with your mood, or, as we have all been urged to become aware of, a malevolent virus. The recent COVID-19 pandemic made explicit what many, of course, already knew: The relationship between people and their surrounding world is all but clear-cut and bounded (cf. Manning, 2013; Smith, 2012; Strathern, 1988). Air transgresses such distinctions, and sequences of touch from door handle to nose may transfer virus from one person to another. COVID-19, in that sense, enforced a need to rethink how people relate to their surroundings.

One tool to curb contamination during the pandemic was to install social distance regulations, either through curfews, limiting mobility, or enhancing distance in social interaction to one or two metres. This left a mark on cities across the world. In Denmark and Norway (Sweden had a somewhat different strategy with fewer regulations), one of the striking things about cities during the pandemic was that with recurring lockdowns, atmospheres changed. Parts of the city became empty, more silent, and more unfamiliar as tourists stayed away, people worked from home, and restaurants and shops shut their doors to customers. To some people, this emptiness became an attraction, as it had a certain intriguing, apocalyptic feel to it. One thing is the powerful images in the media of famous places in unfamiliar conditions, another thing is to feel the sense of solitude that can arise from actually being in such a place in person.

For example, Thomas, a man in his 30s living in Copenhagen, explained how he needed to get used to this new sense of the city: 'Copenhagen has become more of a ghost town. And that, of course, does something to, well,

DOI: 10.4324/9781003379188-2

the atmosphere, one might say.' What he describes here is strikingly similar to what has been termed *kenopsia* (from the Greek *kenosis*, meaning emptiness, and *opsia*, meaning seeing) (Koenig, 2021, pp. 185, 186) and described in the *Dictionary of Obscure Sorrows* as,

> The eerie, forlorn atmosphere of a place that's usually bustling with people but is now abandoned and quiet – a school hallway in the evening, an unlit office on a weekend, vacant fairgrounds – an emotional afterimage that makes it seem not just empty but hyper-empty with a total population in the negative, who are so conspicuously absent they glow like neon signs.[1]

Despite the somewhat ocularcentric notion that this is a phenomenon that is *seen*, rather than *felt* through all of the senses, the pandemic left cities in radically new affective conditions. Put simply, cities had their atmospheres reconfigured as people used them in different ways than usual. This new, empty condition imposed itself upon people in different ways, seizing them as something simultaneously subject to recognition as well as surprise, albeit often in ways that left them intrigued rather than unsettled. Indeed, many of the people we met noted that they had actually made trips into city centres from their residential neighbourhoods in order to experience this kind of emptiness now that tourists were largely gone and people had no errands there, as all the shops were closed. They wanted to experience the familiar having become unfamiliar. A term like *kenopsia* is not exclusive to the pandemic situation, of course, but covers a particular kind of atmosphere resting on the feeling that something is missing and, with that, a preconfigured idea that something *should* be there.

Figure 2.2 An empty place: Svend Aukens Plads in Copenhagen.
Photo by Mikkel Bille.

We use this example of the changing urban atmospheres during the COVID-19 pandemic to show that not only are physical objects and human bodies porous, with substances, bacteria, and viruses moving across boundaries and lingering in the air or on surfaces, but so too are the affective qualities of people, places, and things. In other words, emotions and feelings are not just inner phenomena – mental constructs – but something thoroughly embedded in both human bodies and affective environments (see also Casey, 2022; Davidson et al., 2005; Smith et al., 2009). This view of the ephemerality of the city and its atmospheres as porous is reminiscent of what Walter Benjamin and Asja Lacis observed in Naples in the 1920s: 'As porous as this stone is the architecture. Building and action interpenetrate in the courtyards, arcades, and stairways. In everything they preserve the scope to become a theatre of new, unforeseen constellations' (1925, pp. 165, 166; see also Thelle & Bille, 2020; Wolfrum et al., 2018). Vividly describing the rhythm and colour of life in the city, where Sunday's religious festivals filled the street only to make way for Monday's everyday public life, Benjamin and Lacis come to conclude that porosity 'is the inexhaustible law of the life in this city, reappearing everywhere' (1925, p. 168). In this version of the city, everything interpenetrates: Interiors and exteriors, day and night, private and public. Benjamin and Lacis present porosity as a foundational condition of not just architecture, but of the city and its life as such. It is a condition that facilitates 'unforeseen constellations'. In other words, porosity is also what facilitates and comes to characterise the relationships that shape and are shaped by the city. As Doreen Massey was to write much later, places are 'not so much bounded areas as open and porous networks of social relations' (1994, p. 121). Even in the most mundane encounters, whether in the Naples of the 1920s or the Copenhagen of the 2020s, porosity and social relations shape cities.

The implication of this spatial, biological, and affective porosity is the questioning of clear separations between subject and object and a necessary recognition of the foundational relationality of life. Following such a view, the relations that shape the world may seem endless. Architecture, for instance, is made up of heaps of tangible elements: Doors, bricks, pipes, wires, slates. Architecture is likewise a central part of urban spaces that are also made up of roads, trees, benches, sewers, lamp posts, and pedestrian walkways. Additionally, we might include intangible digital signals travelling through Wi-Fi, Bluetooth, etc., as central parts of the urban infrastructure as well. Humans, meanwhile, are made up of organs, tissues, cells, etc. But human life is inherently more than the sum of body parts. It is shaped through the *relations* between people, places, and things, as well as marked by the porosity of these. Following Erin Manning, '[r]elation folds experience into it such that what emerges is always more than the sum of its parts' (2013, p. 2). The affective relations we are concerned with are forged in atmospheres, or as Böhme puts it, atmospheres themselves 'are something *between* subject and object. They are not something relational, but the relation in itself' (2001, p. 54, original italics, our translation).

This chapter is about ways people shape or are caught up in atmospheric relations to places and other people. It is about the senses of being alone and together in cities, and about finding oneself attuned *by* and *with* others. Cities and their people attune urban dwellers and visitors, and as people move around in them, they do so while being in – or finding themselves confronted with – certain atmospheres. They may thus find places for sorrow, contemplation, intimacy, happiness, conviviality, excitement, or vibrancy. This is the crux of our argument in this chapter: People are always already attuned, but this attunement is subject to constant change along varying relational nexuses between people and spaces. While people might know certain places by their atmospheres – a cosy street or a serene park – and chase them down, hoping to achieve a certain affective experience, there is no solid fixity to these atmospheres as they are experienced and co-produced by people. Just as there is a materiality to the atmospheric city, as we will explore in Chapter 3, there is also a social relationality to it. One consequence of this is the necessity of considering that the quality of a space might not simply be that it gathers people and offers possibilities for social interaction, but also that it allows for people to seek it out exactly because of its lack of other people.

Chasing resonance

'The sight of people attracts still other people' – this observation was made by Jane Jacobs (1961, p. 47) and has shaped much of twenty-first-century urban design. In recent years, in particular, the importance of cities having so-called social infrastructure (such as libraries, lidos, parks, sports fields, skateboard parks, etc.) has been highlighted with reference to the ability of such infrastructure to establish social connections and encounters, thereby potentially strengthening a sense of community and trust. The latter even extended as far as to the level of the state to the extent that trust is established in the provision and maintenance of the facilities (Amin, 2006; Klingenberg, 2018; Latham & Layton, 2019). When Jacobs formulated her perspective in the early 1960s, she did so as part of a critique of how city planners and architects operated from the premise that people seek out emptiness, order, and quietude. In contrast, she argued for the importance of a lively street, which 'always has both its users and its watchers' (1961, p. 47). This liveliness is – or should be – at the heart of constructing cities for people with places to move, sit, and meet. As Gehl later argued, '[b]eing among others, seeing and hearing others, receiving impulses from others, imply positive experiences, alternatives to being alone. One is not necessarily with a specific person, but one is, nevertheless, with others' (2011, p. 15). In the present day, *who* is present in cities and *what* they do in them are issues that we take to be naturally important to urban designers, but how can we qualify how it *feels* to be in cities as social spheres?

In Stockholm, we met Lars, a 56-year-old insurance salesman, who, like many others we spoke to, highlighted the sound of the city when describing his experience of it. He liked the 'life', 'pulse', and 'speed' of the city that the sounds around him brought forth. At the same time, he explained how the level of sound – or noise – differs dramatically depending on one's location within the city. Lars noted how in some areas there is 'this stillness where you can be alone, where it is really quiet. Spaces that are beautiful in that way' (see also Gehl, 2011, pp. 166–170). What he liked the most about the city, meanwhile, was the combination of quiet places and the vibrancy of fast-paced and loud places, each accessible from the other in just a few minutes' walk. Indeed, in moving through the city, he would 'choose the atmosphere' to go to. As he told us, he would intentionally allow for the atmosphere of a particular place to seep in and influence his thoughts and feelings and would effectively *chase* atmospheres that resonated with him at a given time, whether he was looking for calmness or vibrancy. This active selection of location involved him going into the city already attuned, but also already aware of where he might go to find the kind of atmosphere he desired (see Sumartojo, 2022, pp. 116–118).

Figure 2.3 A quiet place in the city: Hammarby Kaj in Stockholm.
Photo by Siri Schwabe.

As he continued to describe the qualities of different places, Lars specifically mentioned quiet spots, where he would go when he just wanted to 'be in his own thoughts' and would go searching for quietness to mirror his state and wish for focused introspection. But at other times, he told us, he might find himself searching for a vibrant area to have an afternoon glass of wine around other people or looking for a playground to take his children to so that he could experience *their* joy at playing. Following the examples of Lars and many other of our participants, city life involves ways in which people's surroundings *resonate* with how they feel or want to feel. The atmospheric city, in that sense, is a city of multiplicity that harbours both loud and quiet places – these adjectives understood both in the narrow, sonic sense and as wider-reaching metaphors that speak to certain atmospheres – and that can facilitate different feelings in and of place. One could argue that the success of an urban space could thus also be judged by its ability to attract people who just want to sit there all alone, and not merely by its potential to attract evermore people.

However, loud and quiet places cannot always be expected to be loud or quiet. *What* atmosphere is *de facto* experienced in them relies entirely on the conglomeration of relations between attuned beings and environmental elements coming together at a specific time. If Lars had gone to his favourite spot at a local square for a glass of wine at a very different time, he might not have found the pleasurable atmosphere he would usually go there to find, nor would he expect to. If he had come during the period after closing hours but before the city usually rises for a new day, the square would most probably have been marked by dark corners and more or less emptied of human presences. He might have run into a couple of insistent party-goers intent on keeping the night going beyond closing time, or perhaps a rat scurrying across the pavement in search of food, taking advantage of the relative lack of disturbing factors to such a project. If Lars had gone at 10 o'clock on a Wednesday morning instead, he would have encountered a different scene entirely, even if the pavement and building facades were the same; perhaps busy with mid-morning coffee drinkers, tourists on tour, and retired dog walkers, but presumably very little wine-drinking.

It is easy to imagine how different the atmospheres would be in these scenarios, despite the space and its material markers remaining relatively unchanged. Repeat the thought experiment to include different seasons, weather types, events, or the advent of a pandemic and the atmosphere keeps changing accordingly. Thus, while there may be limits to the variability and speed by which it changes, the atmospheric expression of a specific space may differ tremendously over the course of a day or year. While we posit that the non-human elements of the urban environment play a crucial role in the constitution of atmospheric cities, it is also clear that human presences and absences are central to how cities and their particular streets, parks, nooks, and crannies feel. Not only do they change character according to the presence or absence of people, they also change character according to what *kind* of people are doing *what*, and what psychological and physiological dispositions they bring with them into the space; how they are attuned. While many things are shared by humans, diversity cannot be underestimated either, in particular, with regard to social and bodily capacities and positionalities.

The process of attunement that Lars describes tends to revolve around a search for (but not necessarily always the achievement of) resonance. Shannon Mattern (2020) suggests that, in 'a world defined by climate crisis, surveillance capitalism, and the periodic collapse of global health, we need to think as much about a city's *resonance* as we do its resilience and livability'. The term has certainly become popular lately, not least with the work of Hartmut Rosa (2019), to whom resonance is a relational mode involving mutual response. It is what happens when people encounter someone or something that strikes a chord. Significantly, in Rosa's view, these sorts of encounters cannot be programmed or planned ahead with any promise of success; indeed, they require a certain level of surprise and lack of control to come about. In cities and elsewhere, 'when or where resonance will happen is unpredictable' (Rosa, 2019, p. viii). When it comes to experiencing resonance, 'we are not outside or indifferent to our surroundings, rather we vibrate with, in and through them' (Thibaud, 2015, p. 41; See also Sumartojo & Pink, 2019). As was apparent among the many people we talked to, when they are part of atmospheres that resonate or dissonate with their attunement, they become integrated into the emergence of the world they are in – or seek to avoid – to a great extent because of the porosity and interrelatedness of people. The example of Lars illustrates how resonance is something that can be aimed at, sought out, and achieved on the basis of existing perceptions of particular places, but the outcome cannot be taken for granted; it hinges on people.

Solitude in the atmospheric city

The renowned Danish artist Vilhelm Hammershøi's (1864–1916) paintings of Copenhagen are characterised by a pastel-coloured, quiet, calming solitude. In his rendition of the royal palace at Amalienborg Square (1896), no one can be seen through the windows as they reflect the sunlight and only vaguely reveal the white curtains that frame them from within, and the square itself is void of movement. No people – and not even a single pigeon – have been allowed to enter the frame. The only thing reminiscent of life is the statue of King Frederick V, but that too remains a lifeless, grey form. Hammershøi's work with light flattens but also opens up the space; there is no harshness, no unsettling dark corners; the scene is easily accessible and there is no sense of potential disruption.

Figure 2.4 Amalienborg Plads by Vilhelm Hammershøi (1896).

As is conveyed in Hammerhøi's work, people do not need to be around other people to feel an atmosphere. Someone may visit a magnificent landscape – an open marshland or an enveloping, lush forest – devoid of other human presences and nonetheless be completely taken in by it, perhaps precisely because they have it to themselves with no one around to distract them from the immediacy of what surrounds them. This feeling is not exclusive to environments with no obvious trace of humans. Sometimes, even densely built cities may appear still and quiet as people find themselves alone in them late at night, or during a pandemic when streets are empty and hushed. Just as the presence of large numbers of people does not by definition create a positively perceived urban space, so too can being alone in such an urban space be both comforting and eerie. The emotional experience of solitude very much depends on the situation; the time of day, the specific design and stories of the place in question, as well as one's relationship to it, the meteorological conditions, and one's disposition at that moment. Indeed, it hinges on one's attunement, one's particular mode of presence. People may in essence resonate with their surroundings by being in solitude, or they may experience a disruptive sense of dissonance that creeps in as solitude transforms into loneliness or an urgent feeling of vulnerability.

Early on during the pandemic, COVID-19 lockdowns had meant a stark reduction in traffic through the usually bustling district of Vesterbro in Copenhagen. It had become much quieter, almost coming to a standstill. During this time, we spoke to Thomas, whom we introduced at the beginning of the chapter. Despite the relative stillness of the neighbourhood, Thomas told us, it still somehow felt lively. This, he explained, had to do with something more profound than the mere presence or absence of people in public space. Being interviewed over the phone while walking through the neighbourhood on a cool spring evening, what he described was more like a vibrancy almost embedded in the fabric of the neighbourhood, a sense of 'life' despite the relative absence of people. Accompanied by the sound of the impact of his feet on the pavement and a few sniffles along the way as his nose responded to the weather and his movement, Thomas described how he had gone for a late-night walk recently because he was feeling restless at home. During this walk, he reflected on how his surroundings still felt alive, despite the streets being entirely emptied of people. Even while having the public spaces of the neighbourhood completely to himself, he had sensed an atmosphere through the vague sense of life in the buildings around him. As he explained, it was the fact that he *knew* that the area was usually full of life that had left him sensing a continued liveliness as he simply noted that, it being late at night, this life had just gone still and moved indoors, only remaining vaguely visible from the light spreading from the windows. His previous experiences and knowledge – both rational and embodied – that there was life in the area despite its temporary invisibility allowed him to make presumptions that, in turn, shaped his overall experience of moving through the streets alone. He was alone but did not feel alone.

The solitude represented in the works of Hammerhøi is not unlike what Thomas experienced in Vesterbro; the calmness of a quiet public space is simultaneously marked by the sense that there has been, and still is, life here. Elsewhere in the city, we often met people who would search for similar experiences of resonance in spaces of solitude, although that of course did not mean that they necessarily achieved it. For instance, the so-called Black Square, which forms part of the renowned Superkilen public space in Copenhagen, is a place designed with an abundance of spotlights and material elements put in place for people to do things, including playing, eating, and relaxing. Besides a play area centred on a large structure in the shape of an octopus and a small fountain nearby, there are chess tables, barbecues, and plenty of seating. Yet, it was on the top of a dark hill, mostly used for exploring the view or tumbling down at times much warmer and lighter than on this October evening, that we found a solitary man sitting on the ground under a tree near a user-created dirt path in the late evening, smoking a cigarette while looking at his phone.

Facing away from the square itself, he was both hidden from view and had no view to speak of himself, aside from the bright screen of his phone and the apartments in the distance. This was his preferred spot and was a place of privacy, almost secrecy, he told us as we interrupted his moment alone. He came there almost every evening, he said, just to get away from the noise of the family in the apartment next door, smoke a cigarette, and just sense the quietness. There were no design features like benches or light art immediately around him; nothing but a small hollow in the ground, where he could sit somewhat comfortably on the grass. Avoiding the lit-up and otherwise highly programmed parts of the square, he had simply come to sit in a dark spot on the cold ground in search of quietude. Just a few metres away, down the other side of the hill, there was a square potentially bustling with life and brimming with design features to accommodate activities, but he had opted for something else; for a comforting solitude otherwise difficult to find. For this man, the atmosphere he found in this particular place, not too far from his home, spurred on a sense of quiet comfort; a sense of calm contrasting with the noise that arises from life in the city's old apartment buildings, where people live in flats with walls that often let sounds drift fairly easily between them. That this was an experience of comfort for him cannot be understood simply by focusing on the space itself, however. Indeed, many other people might have felt unsafe being alone there, or at least considered the lack of sounds – and people emitting them – as eerie or somewhat unsettling.

The atmosphere experienced by the man on this and other evenings of quiet solitude relied to a great extent on the social circumstances of his presence. More specifically, it relied on his familiarity with the space and further surroundings, not least stemming from the fact that he was very close to home. Similarly to what Thomas described in Vesterbro, this man experienced a quietude that was, however, not lifeless. In that regard, looking at the city from an atmospheric perspective, *life* is not a matter of the number of people present or the advent of an intensely full sensory experience, but about how people find themselves situated as part of particular atmospheres. Places that might appear lifeless as seen from afar are not necessarily experienced as such on site. In addition to that point, developing what we call life-oriented cities does not necessitate facilitating the coming together of many people at any given time in any given public space. Sometimes less can indeed be more, if we follow the many instances of people we talked to seeking out, and finding, places for solitude to resonate with their state at that particular time.

Urban spaces often include vastly different atmospheres and sensations within their disparate corners or extending beyond them. As architecture critic Karsten Ifversen notes with reference to the pier Kvæsthusmolen in Copenhagen, 'at one end there are high-intensity experiential spaces that are all about promoting various community functions. At the other end of the pier is an almost existential space where it's just you and the sky and the harbour' (2019, p. 139). It is not that there are clear tangible borders or boundaries here. Rather, there is an extended atmospheric space slowly turning from intensity towards calmness. That sense of calmness is a rare quality in urban space, which otherwise is often charged with the noise of cars, trucks, bicycles, helicopters, and shouting, but in which people also need 'spaces to seek out when you're heart-broken' (2019, p. 139). Focusing on attracting *more* people who carry out *more* activities to places involves the risk of losing this particular quality of city life as such. As Gehl himself occasionally mentions, there may be a need to 'ensure a more even distribution of city activities over larger sections of the city, or to establish peaceful, quiet spaces as supplements to the more lively ones' (2011, p. 81).

Marina Peterson similarly notes that '[n]oise is atmospheric. Palpable in its sensation, noise is nonetheless ephemeral and indefinite; falling away as both sound and category, it proliferates into an array of atmospheric forms. Attuning toward noise is thus also an attuning toward the atmospheric' (2021, p. 4). Following the examples above, being alone in a space marked by the relative absence of people can bring about experiences of comforting quietude, whether the streets of the neighbourhood have been laid bare or traffic is steadily moving not too far away. These experiences are not merely contingent on the presence or absence of noise, of course, but often rest upon a multisensory perception of quiet that may have only little to do with decibel levels. As Peterson (2021) compellingly argues, noise is indeed deeply atmospheric; it is both something sensed and something through which people make sense of their surroundings.

What the examples described here all in different ways illustrate is that life in the atmospheric city is shaped by the presence or absence of people as well as what these people do and how, but also at least as much by people's own attunement, allowing them to experience places in certain ways. Meanwhile, as we will return to later in this chapter, attunement also involves willingness and work to tune into the urban environment. In that sense, atmospheres rely both on the various ways in which people are present or absent and on the particularities of their engagement with these surroundings.

Being alone together

Of course, one does not necessarily have to be entirely alone to experience quietude and solitude. As Teju Cole's New Yorker protagonist laments in *Open City* (2011, p. 6):

> Walking through busy parts of town meant I laid eyes on more people, hundreds more, thousands even, than I was accustomed to seeing in the course of a day, but the impress of these countless faces did nothing to assuage my feelings of isolation; if anything, it intensified them.

At the same time, on his walks through the city, neighbourhoods make distinct impressions on him as they each 'appeared to be made of a different substance, each seemed to have a different air pressure, a different psychic weight' (Cole, 2011, p. 7). Rather than the sociality of atmospheres being simply about the number of people around, it is about the qualities of the overall feeling of a multifaceted human presence (made up of a number of individual presences) that make the surroundings appear to those present in a particular way. Of course, as Cole suggests, this human presence interplays with all the non-human elements that shape cities to make for neighbourhoods that have distinctive atmospheres that sometimes feel isolating, with the *alone* in *alone together* accentuated.

In either case, people are not necessarily together *with* anyone in particular when spending time in the city, but are nonetheless co-present with others in one form or another most of the time (cf. Gehl, 2011, p. 15). This facet of urban life is sometimes sought after rather than deplored. For instance, we met people who liked where they lived because they experienced their immediate surroundings as being *close* to the city, yet not exactly *in* it. Their homes might have been located centrally but on a side street to a busy square, or vertically located *above* the street, thus offering the impression of being on the margins of the city proper and allowing them a feeling of being alone together with others, with the *together* now emphasised (cf. Bille, 2015).

Similar experiences of being on the periphery, of looking *at* the city not so much from within it as from a vantage point somehow beyond it, sometimes extend to public or semi-public spaces. For instance, the Oslo Opera House offers an example of public architecture that is both connected to its surroundings and simultaneously quite separate, not least from a sensory perspective (see Payne-Frank, 2022; Payne-Frank & Schwabe, 2022). An extremely high-profile architectural project, the Oslo Opera House was finished in 2008 and has since become a local landmark as well as a highly frequented space. Now flanked by the large and well-used Deichman Bjørvika public library toward the city and the Munch art museum further out toward the fjord, the Opera House has garnered much attention due to its design, whose most remarkable feature is the white marble roof that slants down toward the water and provides 38,000 square metres of traversable space that is open to the public around the clock.

36 *Attuned relations*

The roof is busy year-round with sightseeing tourists, local walkers, dancers looking for an open rehearsal space, people on electric scooters testing the slope, and photographers seeking out a landscape motif or a sleek background to their portraits. Meanwhile, the foyer space of the building, accessible through revolving doors at the city-facing side, is also open to visitors most of the time. While not as popular as the roof, this indoor space is almost never entirely empty – even between shows – as people come in to sit at the small café, relax, and perhaps warm up a bit after a wind-swept walk on the roof on the benches along the interior oak-clad wall, visit the small gift shop, or use the public toilets.

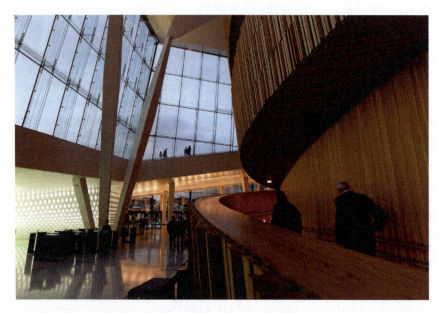

Figure 2.5 The Oslo Opera House foyer.
Photo by Jeremy Payne-Frank.

From this space, in particular, the city is visible but tinted and muffled; through the large glass windows to the front, people sitting on the wool-covered benches can observe the human movements on the near part of the roof as well as take in the sight of the birds and clouds that move along further above. Looking to the right from the same spot, people can follow the street-level traffic, see visitors approaching the building, and trace the routes of the big red buses passing on the road toward the library. As one visitor said: 'I like that it's so open, the windows giving light… It's not walled in, so you can see the other buildings and the water. You can see the traffic lights.'

However, while the city is thus present by sight, it is largely absent through sound, smell, and other sensory impressions. In other words, the life of the city is *filtered* by the glass, marble, and other materials that make up the physical border of the building. Yet, importantly, the city is not absent; it is simply experienced from a vantage point whose peripheral qualities are the result of relative distance and use of materials. This sense of attached periphery is multisensory and experienced through various combinations of sensory impressions, not least the muffled sounds in the background and the refractions of lights coming from a distance.

In the context of social gatherings, this sense of *peripheral attachment* – of being connected to what is going on, albeit from a distance – has been described as 'midding' (Koenig, 2021, p. 85):

> Feeling the tranquil pleasure of being near a gathering but not quite in it – hovering on the perimeter of a campfire, chatting outside a party while others dance inside, resting your head in the backseat of a car listening to friends chatting up front – feeling blissfully invisible yet still fully included, safe in the knowledge that everyone is together and everyone is okay, with all the thrill of being there without the burden of having to be.

Another location in which we encountered this phenomenon was at Israels Plads, a popular square located in central Copenhagen. The space features large rows of stairs on each end from where one can overlook the activities below – people playing basketball, kids skateboarding, neighbours out for a stroll – or eat an informal lunch, perhaps from one of the food stalls at the market across the road. The entire plaza has been elevated slightly in relation to the surrounding streets, leading the designers behind it to refer to it as 'Copenhagen's biggest urban carpet' (Cobe, 2018, p. 73). Indeed, while for many years in the mid-to-late twentieth-century Israels Plads functioned mainly as a large parking lot, the cars have now moved to underground facilities, leaving the square to 'hover over the many cars that used to dominate Israels Plads (…). The cars are literally swept under the rug of the new urban living room' (Cobe, 2018, p. 73).

The characterisation of the plaza as an 'urban living room' is telling of the designers' intention to make this a comfortable and lively space where people might feel at home. And, indeed, some of the people we talked to who lived in the nearby apartments, seem to treat it as a space resembling an extension of home. This is a vibrant place, often brimming with school children from the nearby school, people passing by, and, on sunny days in summer, visitors sunbathing on the stairs while waiting for their companions to bring over beverages or food from the nearby market. People sometimes come around to watch the sky as the sun sets. Others enjoy coming around as it rises in the early morning. One person commented on her habit of coming down to sit on the steps early in the morning just to watch people and witness the city coming alive, explaining how 'it is peaceful at this time'.[2]

One of our participants, Jonas, had passed the square many times, but he proclaimed that the time we met him was indeed the first time he had actually decided to sit down and stay there a while. He was out on a morning stroll, walking the city to contemplate his current life situation, and had gone to a

38 *Attuned relations*

Figure 2.6 Israels Plads, Copenhagen.
Photo by Siri Schwabe.

popular spot nearby, namely a bridge leading from central Copenhagen to the neighbourhood of Nørrebro. Yet, as he arrived there he felt that the 'morning-busy' and 'pulsating' atmosphere became too much. 'I was filled up all of a sudden and thought I had better walk down to Israels Plads' in order to get to a quieter place. As he described it, Israels Plads is a 'philosopher's place' that provides 'calm for contemplation, at least early in the morning'. Jonas was 61 years old and in the midst of a life crisis. He had lost his job and his home and was temporarily living in a hotel from where he had to visit the job centre and union located close to the square. But today was different. He did not have to go to those places and came instead to the square. Noting how the children from the nearby school would come out to the square for breaks, Jonas described how they add a sense of 'peaceful life' as they play while supervising adults observe them from the periphery. 'When you reach my age, children are life affirming', as he noted, continuing: 'They make you reflect upon your own life: "have I lived the right life, what kind of hopes and dreams did I have as a child?"'

To Jonas, in addition to being a peaceful and quiet space facilitating contemplation, the square seemed 'open', offering, in particular, an 'openness' towards the adjacent building facades. There was nowhere to hide in the square, especially not where he was sitting, yet he did not feel observed or exposed in any way. He felt able to simply sit there, relaxed, and, as he related,

> have space to let my thoughts run freely. And if I don't know what to do I can just sit and observe people around me. Where I am in life right now, it's actually quite pleasant to just sit and observe other people,

He continued reflecting on the sensation the square offered him and concluded that 'this square gives me the feeling of freedom'. While there are people who just pass by, as he usually does himself, there are also people who 'are just here and have the time to be here, including myself'.

What we take from Jonas' example is that his experience on that morning was shaped by a determined search for an atmosphere and a space where he could feel calm and free to contemplate his own life by being on the periphery of the space, observing the children playing and all the other people moving in and through the square, all with their own lives and concerns. As we talked to some of the other users, it became clear that the design of the square offers them a feeling of being part of an atmosphere but being so at a distance. The stairs, in particular, indeed act as a vantage point allowing people to be part of what some called an 'effortless atmosphere'. Another person we spoke to similarly reflected on the atmosphere on the square: 'You're *together* in a way, without even knowing each other.'

The point with this example is not to essentialise the character of this specific square or how people experience it, although the design does offer the potential for a sensation that might be termed *peripheral attachment*, especially for those who take in the square from the stairs on either end. Indeed, some respondents also felt excluded by the minimalist design and the highly commercialised and – some would say – overpriced food hall that had replaced the traditional open-air fruit and vegetable market that used to occupy the adjacent space. However, what could be interpreted as a sense of peripheral attachment or 'midding' was a common response, also from residents of the apartments overlooking the square that we spoke to. They noted how they experienced a sense of informality and relative anonymity at the square, even from their homes, while they enjoyed the abundance of sound as they followed the rhythms of people moving around the space, including the school children using the square as a space of play on their breaks. These experiences of being on the periphery highlight the blurred boundaries of atmospheres in the physical border zones between private and public. If we take what they experience there as a type of peripheral attachment, it is not about solitude or loneliness. It is the comforting, yet subtle, relation between participation and periphery. In that sense, it is different from the withdrawal one sees when people cut off their awareness of their surroundings, for instance, getting sucked into a world of text, images, and sounds from a phone, retreating from the here and now (see Bull, 2007; Leder, 1990). Peripheral attachment, on the other hand, involves a withdrawal from an active engagement with others, yet not a withdrawal from being attentive (for more on this argument see also Bille & Hauge, 2022). At Israels Plads as elsewhere, atmospheres move across physical perimeters, cross streets, and flow in and out of windows, up and down stairs, highlighting the porosity that undergirds life in atmospheric cities.

What people's experiences with withdrawal in the presence of other people tell us is that urban dwellers do not necessarily have to be directly involved with those around them to feel part of a shared space, to experience and take in something of an atmosphere that is, at least in part, formed through social relations. Indeed, being at a distance – and perhaps experiencing a sense of attachment on the periphery of a situation – allows for certain kinds of attunement, which, in turn, often bring about a subtle but nonetheless deeply felt experience of resonance. Even from the periphery, people vibrate with the social situations unfolding around them.

Dissonant encounters

Being alone among others can of course also invoke experiences of dissonance, like that which Cole's protagonist experiences in New York, and so can finding oneself alone in a dark public space such as the Black Square in Copenhagen. Solitude often does not imply a sense of comforting quiet; rather, to many other people – especially those occupying more vulnerable positions relative to their surroundings – looking for calm in a dark spot away from other people would be both a risky and ultimately fruitless endeavour, as the calmness would most likely remain evasive and instead be replaced with anxiety, nervousness, or even fear. Being entirely alone or simply being on the periphery of a space filled with people can entail discomfiting experiences of dissonance when people find themselves feeling out of tune out of tune with their environment. While numbers may matter – whether there are zero, five, or 85 people around – a key to understanding atmospheric urban spaces is the relational character of what and who are present or absent. With that, atmospheres evade fixity since the constellations of people, things, and phenomena that constitute them, fluctuate. People's attunement is also an ever-changing process that involves sensing different places differently, tuning into atmospheres as they shift.

In Oslo, several areas along the harbourfront have undergone development in recent years and decades. Not far from the Oslo Opera House and the surrounding Bjørvika area, we find another neighbourhood on the fjord, Aker Brygge. Originally home to Oslo's foremost shipyard, which closed in 1982, the area was inaugurated as a modern-day neighbourhood in 1986 and has since then undergone a process of further redevelopment in 2011–2014.[3] With a mix of housing, eateries, a conference centre, and a vast array of shops, Aker Brygge is a lively district, not least along the boardwalk that lines the eastern-facing part of the neighbourhood and that features a number of high-end restaurants. But it is also a district with a reputation for catering specifically to businesspeople and wealthy consumers.

Attuned relations 41

The perception of Aker Brygge as exclusive and as a place that is unwelcoming to people who do not fit in with a certain clientele seems to come across to some locals in both thought and feeling. During an interview on-site, Anna, a young woman who had grown up nearby, described it as having an 'unpersonality'. Anna's way of describing the area pointed to a lack of character so distinct that the people we spoke to about it often had a difficult time conveying any particular sense of it at all, besides not liking it very much and not feeling inclined to spend much time there. Despite it often being quite busy, those who move through the space do not seem to dwell there for long; as one person put it, using the term 'atmosphere' to denote a positive experience, 'the people walking here are always going somewhere, and that contributes to this non-atmosphere'. To Anna, the rhythm of the neighbourhood was somehow 'off', and she was struggling to connect with it. She described the view of the fjord from the waterfront as one of the loveliest Oslo had to offer. Yet, she remarked, people never seemed to stop to look, evincing an inattention that made her uneasy. Between her own attunement and other people's ways of inhabiting Aker Brygge, there was a dissonance. Moreover, there was something uninviting about the built environment of the area itself to her. Instead of filling the harbourfront up with outdoor seating for restaurants and bars, she added, those behind the development of the area might have kept the fjord-facing promenade open and more easily accessible in order to have 'taken advantage of what is so beautiful about [this] nature instead of forcing us to walk around among impersonal apartment blocks where only rich people, who I don't feel any connection with, live'.

Figure 2.7 Aker Brygge, Oslo.
Photo by Jeremy Payne-Frank.

To someone like Anna, there are layers to what she described as a lack of connection to this place. There is the felt dissonance – the feeling of being out of tune, off-beat – but there is also the influence of certain ideas that seem to be present both in Anna's perception and in wider public discourse, namely that Aker Brygge is a neighbourhood chiefly for the affluent. In that sense, the feel of a place such as this may actually become a sensory marker of social division. Anna entered the space of the neighbourhood already marked by established ideas of it. She was in that sense already attuned; already prepared to feel out of place. She was perhaps unable to resonate with her surroundings in Aker Brygge, but, additionally and importantly, she was also very reluctant to do so; she had to some extent already decided beforehand that this atmosphere was not for her. Also, in this case, there is a porosity at play that works at several levels: People's distinct presences – their dressing style and bodily dispositions that play a role in shaping them – intermix and make for an atmosphere that, likewise, spills over across geographical distances, travelling with Anna and others as they bring their impression with them. Similarly, Anna's attunement was already, upon her arrival, marked by the atmospheres she came from as well as her expectations of the atmosphere she would find. In other words, just as humans are porous and always interrelating, so are atmospheres interpenetrated by other atmospheres forming alongside or adjacent to them.

In Copenhagen, Thomas juxtaposed his night-time local walks with experiences of being alone in the public spaces of other parts of the city. In contrast to Vesterbro – an old, inner-city district – Thomas had also visited more newly built neighbourhoods on the outskirts of the city and described his sense that these were like 'ghost towns' after dark, leaving him feeling as if they did not 'invite' him to be there. To Thomas, these areas presented him with what he called a 'dead atmosphere' marked by a lack of life and reflecting what he considered 'failed' urban planning where everything still 'looks so new' and lacks the 'patina' of the places where he enjoyed spending his time. These 'uninviting' spaces – not unlike the Aker Brygge of Anna's experience – appear to people like Thomas and many others as uncanny or, if we translate Freud's term *unheimlich* more accurately, unhomely, elucidating the discomfort of discovering 'the strange *within* the familiar, the strangely familiar, the familiar as strange' (Fisher, 2016, p. 10, original italics). Areas like these feel, at least to some people who fall outside their target user groups, as if they are just a bit *off*. They might resemble places in which people usually feel a certain connection to their surroundings, but still, despite retaining traces of familiar urban phenomena – residential blocks, street signs, perhaps some scattered greenery – they remain something other than what some people know and recognise as homely.

In both Oslo and Copenhagen, relatively newly developed districts were described as neutral, sterile, deserted, and unwelcoming. Many readers will probably recognise the feeling of lifelessness in urban areas that appear to people as lacking a solid grounding in the city; something 'more' than just architecture as an outer shell (see Pérez-Gómez, 2016). This grounding may be experienced through the presence of old trees, lights in the windows, or the weathering of building materials over time, whether they become dust-stained, rusty, or covered in green blotches from intrusive organisms that thrive in damp conditions. When a new neighbourhood retains a feeling of unfamiliarity and those who move within it are unable to get a clear sense of life – and what that life might be like – in their surroundings, it often leaves them feeling wrapped up in an uncanny, 'dead atmosphere', to stick to Thomas's phrasing.

Put simply, the familiarity of Vesterbro, Thomas's intimate knowledge of it, and not least the fact that he knew what he might expect from it all led to his resonating with it in a situation where the lack of human presences might have resulted in the opposite. Conversely, the strangeness of a newly built neighbourhood that was not quite what Thomas had come to expect from an urban residential area in its lack of 'patina' led him to an experience of dissonance. He simply did not recognise this 'ghost town', whether sensorially or cognitively, as what a Copenhagen neighbourhood *should* be. Like Anna's experience of Aker Brygge in Oslo, Thomas felt out of place here; it did not seem to be a neighbourhood for him, and rather than attempting to attune himself to its atmosphere, to put in the work needed for him to resonate with it, he rejected it and ultimately removed himself from it.

These stories highlight, firstly, how experiences of the uncanny bring about a sense of dissonance, an inability to resonate with one's surroundings because they are simultaneously familiar and strange; there are people but no life, buildings but no sense of atmospheric grounding. Apart from architectural and design differences, these neighbourhoods *look* like the neighbourhoods people know and are made up of the same structural elements. Yet, to some of the people we talked to, they fail to facilitate familiar atmospheres and so appear as strange. Design clearly plays a role in these cases, as does the passing of time; newly built neighbourhoods lack the signs of age and the grounding in space that their older counterparts have. Importantly, this does not mean that people living there do not find a positive atmosphere or resonate. As illustrated in the Introduction to this book, the very narrative of what a place is may cloud what is actually felt when present in it.

Secondly, and importantly, these experiences are very much about attunement as a social phenomenon, and about how porosity marks inevitable and ever-evolving relations between people, places, and atmospheres. Being attuned to a space is not something static and fixed, *a* feeling. It is rather about *feeling* as a verb, or constantly *getting a feel* for spaces and situations that often overlap in experience. It is a process with its ups and downs, where some spaces at some point may invigorate or pull you down, and where small changes matter. It was not uncommon during our fieldwork that we and other members of our research team found ourselves in urban areas that at first did not feel comfortable or safe. Yet, a process of slow attunement to the place most often occurred during which an increasing familiarity rendered the atmosphere well-known, leading it to fall into the background, perhaps be taken for granted, and perhaps even end up being appreciated.

At the same time, the process of attuning to a space can also be marked by a certain level of unwillingness or refusal to tune in, as was the case for both Anna and Thomas as they visited newly developed districts of their home cities. Following their examples and others of our participants, it often takes atmospheric work to gain and retain a sense of place (cf. Ahmed, 2014). Seeing their experiences of dissonance through the lens of atmospheric work opens up for an understanding of attunement as shaped through some measure of choice, whether conscious or subconscious, as well as mundane, everyday practices (see also Bille & Simonsen, 2021; Gherardi, 2017). Thomas and Anna strongly disliked the new districts of Copenhagen and Oslo, respectively. They did not *want* to resonate with them; they did not agree with the politics behind their development, and they did not identify with the people otherwise occupying these spaces. In that sense, the spaces caused them to feel a dissonance, but, conversely, they also refused to tune into them. They were, in Ahmed's (2014) words, 'not in the mood'.

In a broader sense, then, atmospheres are not just something that happens to people as they find themselves in various spaces that speak to them in different ways. An atmosphere is not just a visceral, sensory, or affective phenomenon in that sense. It is also in certain ways a practice: A tuning into or tuning out of (Bille & Simonsen, 2021). In certain ways, it is indeed a kind of work, something that takes effort to achieve and something that can, reversely, be resisted. The presences and absences of other people are crucial to how atmospheres work in that regard, but there is of course much more at play than that. As a social phenomenon, the atmospheric city is also about the ideas, perceptions, and attitudes that people bring with them into a space and into the relations that mark their experience of that space. Also in that sense, atmospheres and atmospheric cities as such are porous, marked by all the stuff that spills over from one body to another and one space to another, while also marked by the experiences that people bring with them as they move from one neighbourhood to another and back again.

Quiet and loud: Crowds and atmospheres

While processes of attunement never cease, people are perhaps never more likely to truly feel part of a shared atmosphere, a movement, than during events in the public spaces of the city. When people get together by the hundreds or even by the thousands during festivals, rallies, or protest marches, something happens to them, and something happens to the city.

In Copenhagen, we spoke to Marianne, a retired woman in her 60s who lived on her own in Nørrebro. Back in 2015, Marianne had attended a large public memorial demonstration outside the French embassy in Copenhagen in the wake of the shooting at the satirical newspaper Charlie Hebdo office, which left 13 people dead. As she told us years later, she had wanted to leave flowers at the scene and thus went with a friend, encountering a sizable gathering as they arrived. Reflecting on this experience, she described in great detail how the space

> was filled with sorrow. It was one of the terrorist attacks that really got to me. There was a speech by the French ambassador, and it was so quiet there. Thousands of people, and it was totally quiet. And when we left it was all silent. The most striking thing was that it was only after we had left and sat somewhere else that we came to talk about how silent it had been.

At first, Marianne and her friend had been unsure of whether or not they should attend the memorial at all because they knew it would be a sorrowful and emotionally challenging event. In the end, however, they were pleased to have gone and surprised that, despite the large number of people present, there was so little pushing among the crowd and rather a sense of absolute quietude. Their experience had been 'atmospheric in a sad way', as she put it, and the affective weight of the experience, she said, had 'forced' them to go to a café and have a glass of wine to regain their calm afterwards. It had been an almost overwhelming experience of an almost reverberating silence.

It is fair to assume that the gathering was not indeed entirely quiet in an absolute sense, but nonetheless, Marianne had been almost overtaken by the pervasive sense of sombre calm that she experienced. Marianne's example is interesting not least because it illuminates how atmospheres that arise when people gather are not always the outwardly intensive events we often come to think of when thinking of crowds; the collective physical outpouring of shared emotions that one may associate with noise and movement. This was no spectacle in the commonsense meaning of the word. However, it was very much an instance of intense attunement; an instance of people responding to each other, to the situation, and to their shared space in a way that allowed Marianne to feel an intense sense of resonance.

The presence of a crowd of people played a significant role in that regard. If she had visited the site alone to place flowers in commemoration, one might assume she would have likewise felt something, but that 'something' would have been different and perhaps even more difficult to articulate. In this case, it was by and large the quiet crowd that instilled in Marianne a powerful sense of a shared atmosphere. This was a mode of being together *as a crowd*, collectively present, and present with a shared purpose, contributing to a collective atmosphere. Sometimes, as in this instance, crowds come together quietly, yet with an atmospheric intensity that is felt as strongly as during loud concerts, vibrant festivals, or riotous protests carried by louder expressions of indignation.

At the same time, when thinking of crowds, we tend to think of large gatherings of people who come together for more spectacular events in the public spaces of cities, often disrupting these spaces in the process. Numerous scholars have dealt with the social workings of such crowds and scrutinised how temporary collectivities arise along with atmospheres in crowded spaces;

indeed, there is a long tradition within sociology and social psychology of attending to what happens when large groups come together. In his work, Christian Borch has traced the development of this field of research, positing 'rational approaches' against scholarship that focuses on the 'collective emotional arousal' of crowds (2013, p. 584). While the latter follows a trajectory of thinking established in the late nineteenth century, the former came about in earnest in the 1960s, when sociologists began to consider how collective endeavours such as protests often carry meaning, reason, and purpose with those involved (Borch, 2013, p. 585). Meanwhile, Borch argues, images of irrational and emotionally driven crowds persist, harking back to the work of Gustave Le Bon, who saw crowds as 'characterized by an impulsive, barbarian, feminine nature and by the incapacity to reason' (2013, p. 587). This juxtaposition of rationality and irrationality in prevalent thinking on crowds provides an interesting background for a discussion of the role of atmosphere in situations where people come together, as well as of the role of large gatherings in the constitution of atmospheric cities. Perhaps the most fruitful perspective in that regard lies somewhere in between the two positions outlined above, where we might consider crowds to be made up of people who are both full of purpose as well as being emotive beings whose engagement with the world is never purely 'rational' (see also Chowdhury & McFarlane, 2021).

Also highly relevant from an atmospheric perspective, Émile Durkheim's concept of *collective effervescence* describes what happens when the energy of a group of people intensifies and takes hold of those present. To Durkheim (1982, p. 56), this effervescence is an outburst of collective emotions, which results from people being together, and

> a product of the actions and reactions which take place between individual consciousnesses; and if each individual consciousness echoes the collective sentiment, it is by virtue of the special energy resident in its collective origin. If all hearts beat in unison, this is not the result of a spontaneous and pre-established harmony but rather because an identical force propels them in the same direction. Each is carried along by all.

This sense of collective effervescence spreads when '[e]very emotion expressed resonates without interference in consciousnesses that are wide open, to external impression, each one echoing the other' (Durkheim, 1912 [1995], pp. 217, 218). If we attend further to the affective qualities of crowds, it has been noted that a certain feeling of joy may be experienced when people are assembled, a 'crowd joy': 'the dominant emotion of face-to-face assembly' where people experience various 'levels of arousal' (Lofland, 1982, pp. 356, 357). Sharing both time and place in this way brings and binds people together in a social exchange that is sensory, often emotional, and that exudes atmosphere. With that, public spaces are spaces for voicing collective politics in ways that often engender intense sensory experiences and are marked by atmospheres that centre on joy, anger, or other effervescent emotions that 'rise up' from a gathering (see also Davidson et al., 2005; Smith et al., 2009).

In Copenhagen, we spoke to Kim, a 26-year-old student, who reflected on a recent event during which people had gathered to protest the continued existence of deportation camps housing refugees. Kim described the excitement of coming together with others to convey their message, to 'say it loud and clear and saying it together in public space so that no one will not hear it'. In continuation of that, they explained to us that

> a good *demo* (demonstration) is a demo where the police stay at a distance and don't get involved unless there's a need for it; and where you can yell out chants and sort of use the street to relay a political message.

There had been excitement, but also a sense of comfort in being able to use the public space in this way. As Kim put it:

> It was a way of using the space of the city for something other than the usual. Instead of people just sitting around drinking lattes or whatever, we somehow said, 'hey, we're just going to interrupt this ordinary, everyday life for a moment because something is going on'.

Taking up space to interrupt the usual workings of the city, and doing so loudly and along with others, had been a way for Kim to experience a sense of agency and ownership while expanding the scope of the possible. Along with fellow protesters, Kim had challenged the usual ways of being in that particular space and thus opened it up to a different atmosphere. On that occasion as well as others, they had taken on a deliberately active role in shaping the atmospheric city, in coproducing a sensory environment geared toward communicating a particular political message that in a sense came to overflow the sensory environment of that particular part of the city while the protest was going on.

One characteristic of crowds such as the one Kim became part of is, of course, that the bigger the gathering, the more unlikely it is for one person to know everyone present. As such, there is always a certain level of foreignness, even to some extent anonymity, as people find themselves surrounded by a jumble of friends and strangers (see also Sendra & Sennett, 2020, pp. 17, 18). In that sense, there are similarities to be found in experiences of being part of a crowd and experiences of peripheral attachment as we discussed above; in both instances, feelings of togetherness and social (or *felt*) distance pervade the atmosphere simultaneously. People are in a shared space along with those that surround them, but that does not necessitate any familiarity or particular attachment to them. Indeed, what shapes these affective fields are the particular qualities of the presences and absences of people, which facilitate people's attunement to the spaces. In both Marianne's and Kim's stories, a shared purpose characterised their coming together in a crowd and taking part in atmospheres of sombre quiet and energetic protest, respectively. With that, the human presences that marked these spaces were of a specific kind; people were present in certain ways, embodying an attunement to match that of the crowd as such.

Conclusion

This chapter has shown how atmospheres are a deeply social phenomenon. However, *how* people attune to spaces and their atmospheres is not necessarily so much contingent on human presences and absences in simple metric terms – as in *how many* people are in a certain place or not – but to a greater extent on the quality and relationality of these presences and absences. Solitude can evoke feelings of eerie discomfort or of quiet comfort as the spaces around people bring them into modes of dissonance or resonance. In turn, being near unknown others can make people feel at ease through a form of peripheral attachment or, reversely, bring a feeling of loneliness or isolation into ever-sharper focus. Finally, being immersed in a crowd with a shared purpose – whether commemoration, celebration, or protest – can bring about sometimes almost surprising experiences of resonance. Put somewhat differently, atmospheres sometimes facilitate not so much a 'feeling-of but a feeling-with' (Manning, 2013, p. 8) – a feeling where the separation between self and world is re-conceptualised and less clearcut. In such instances, people do not just get a feel *of* an atmosphere but feel this atmosphere *with* others (see also Dreyfus, 2012).

The implication of atmospheres as inherently social comes with two important corollaries: First, how atmospheres are experienced highlights the porosity of humans, things, and life as such in cities and elsewhere, and, second, attunement to atmospheres is not simply a passive process but often involves a certain level of activity, work even. To tune oneself into one's surroundings sometimes takes labour, especially when finding oneself in a situation or space that is somehow uncanny, foreign in one sense or another, or simply outside the scope of what has already been established as preferable. Rejecting such attunement to atmospheres is also in that sense a choice; and one that is as significant as its opposite, namely, to decide to put in the effort to 'get in the mood' and accept being part of an atmosphere. Atmospheric work can thus carry a political component, as people sometimes simply refuse to be part of a shared space whose sociopolitical foundations seem inacceptable (cf. Ahmed, 2014). People simply cannot avoid other people and what they do or have done to the environment they find themselves in. Likewise, the notion of atmospheric work hints at an entwinement between attunement and the social context within which people experience atmospheres; it is about distinct presences and absences, but even further, it is about all the baggage people carry with them as social beings, and about the wider societal contexts in which atmospheres arise.

Thus, and importantly, the experiences we have described in this chapter are by no means static. Resonance can happen immediately as people find themselves swept away by an atmosphere, but it can also be based on sensory norms, learned over time as people grow accustomed to certain places and begin to read them in certain ways or take them for granted. The same is true for experiences of dissonance; what might have been a favourite place a decade ago could easily become home to discomfort as people and the places around them change over time. In that sense, both atmospheres and attunement to them are highly dynamic phenomena. Of course, when we in this way think of the atmospheric city as a 'relational environment' (Manning, 2013), we think of an environment through which both human and non-human entities form constellations and connections. In this chapter, we have focused specifically on the relational aspects of how cities – and people with them – are attuned, highlighting the social aspects of atmospheres and beginning to move towards an understanding of the atmospheric city as social. In the following chapter (Chapter 3), we shift our focus to the second of the four themes that guide our thinking on attuned urban life and the atmospheric city, namely the environment.

Notes

1 https://www.dictionaryofobscuresorrows.com/post/27720773573/kenopsia. Accessed 16 May 2022.
2 For further discussion of the square, see Bille and Hauge (2022) and Zerlang (2008).
3 https://www.akerbrygge.no/aker-brygge-den-utrolige-reisen/. Accessed 30 May 2022.

Figure 3.1

3 Embraced by the city
Feeling the urban environment

New technologies allow for radical new ways of shaping our environment. At the old part of Oslo Central Station, Østbanehallen, ÅF Lighting[1] has, for instance, installed their vision for future lighting: Liquid Light. By using sensors that detect movement and heat, as well as respond to sound frequencies, the light installation follows changes in the immediate environment, adapting the lighting accordingly and, with that, ostensibly creating 'a sense of well-being'. The idea behind the installation is that with the 'right' light 'people become alert and feel comfortable in a space when the light harmonises with their task or location'.[2] From photos, the installation makes for a visually imposing and thus highly noticeable presence as a portal one passes through via escalators between Østbanehallen below and the part of the station that leads to the platforms above.

Figure 3.2 Liquid Light in Oslo.
Photo by Jeremy Payne-Frank.

DOI: 10.4324/9781003379188-3

However, despite its size, position, and potential ability to affect the human body, the installation remarkably seemed to escape the attention of the people we met at the station. For instance, on a November afternoon, we met a couple standing on a platform at the top of the escalators, overlooking the light installation with Østbanehallen in the background. They were deeply engaged in conversation and had been standing there for some time when we asked them their thoughts on the installation. Immediately they looked somewhat puzzled and commented that they had actually not noticed the lighting at all, but had simply found themselves right there, so engrossed in their conversation that they barely perceived their surroundings. A few minutes later we spoke to a man visiting the city from Finland. He had in fact noticed the lights but only at the margins of his attention up until the point when we asked him about them. When prompted to consider the installation, he remarked how the lights in the station, and the Liquid Light specifically, seemed to 'guide your way even when it is dark'. Reflecting further on the orientational quality of the lighting, he explained that the Liquid Light seemed to tell him that 'there's something important' in their direction. He had not noticed how the light panels that make up the installation had been subtly changing while he was in front of them, but as he started observing them during our conversation, he noted how it was as if the light 'by changing colour changes mood'. Following from these examples, despite the lack of overt attention of many passersby to the installation, it could be argued that it imbues the space with a certain atmosphere – or 'mood' if we stay with the wording of the man from Finland – prompting some users to dwell in front of it and others to feel subtly drawn to and guided by it, getting the sense that something important might await behind it.

The environment, both material and immaterial, that surrounds urban dwellers – and other users of city spaces – does not always stand out in any particular way. In fact, it most often seems withdrawn from attention. However, it has great bearing on how cities are shaped and experienced nonetheless. It shapes the bodily felt sensory expression of the city (see also Howes & Classen, 2014; Rodaway, 1994; Urry, 2012; Zardini, 2005). To a large degree, much of the history of urban lighting, for instance, showcases a focus on lighting for visibility, leading to brightly lit cities inscribed with all the politics embedded in how light facilitates what is seen and what is not (and by whom Schivelbusch, 1987). Recently, with the new large-scale implementation of LED lighting technologies, we see urban spaces lit in colourful ways through detailed curation and with the intention of making spaces *feel* a certain way. It is not because spectacular lighting was not present earlier. But the places these new technologies are implemented show the expansion of such spectacle. This hints at a more 'scenographic city', curated through technologies. As part of this development, stakeholders in the realm of urban development increasingly talk about places in terms of their atmospheres, where CGI visualisations are highly curated tools employed to evoke specific dreams and affective potentials (Degen & Rose, 2022; Degen et al., 2017; Stenslund,

2021, 2023). This CGI trend differs from earlier architectural modelling practices such as white 3D miniature models, where it was to a greater extent up to the receiver to add their own future expectations and imagine the projected space accordingly. Now it is served for the viewer.

Alex Rhys-Taylor (2014) notes that the sensory experience of the twentieth-century city was marked by both bombastic spectacles of music, laser, fireworks, as well as less obvious multisensory aesthetic interventions in everyday urban life. He writes that it

> does not merely passively reflect the colossal technological, political, social, and economic history of the twentieth-century city. The ways in which urbanites 'made sense' of their environment fed back into the city and impacted on its physical, social, and economic topography.
>
> (2014, pp. 75, 76)

His claim is that the twentieth-century developments are what have fundamentally shaped the urban changes we see today. As he writes, 'for the most part, all of the changes that would unfold in the early twenty-first century were elaborations of those that appeared amidst the breathtaking sensuousity of the urban century before' (2014, p. 76). To a large extent, we agree. However, while the twentieth century marked the advent of 'the lit city', with more and more parts of the city becoming illuminated in order for people to *see* – and in some places to promote consumption and symbolic power through advertisements and making public and private buildings visible – we also suggest that something distinct is happening today. Urban lighting technologies are no longer just implemented to make things *visible*. They are also used to make urbanites *feel* – either quite literally by producing bodily responses, or by staging a *scenographic city* through lighting that imbues even the most mundane areas of the city with carefully curated atmospheres. Liquid light is an example of this emerging trend.

In recent decades, it has become evident that technologies hold the capacity to both optimise energy consumption and make data available for creating new urban orders. Parking sensors inform drivers of parking availability, electronic signs along central thoroughfares in Copenhagen show the number of bicycles passing per day and advertise the city as a biking capital, while real-time air pollution measurements across the world increasingly warn about the silent killers in the air we breathe. Measurements offer a diagnosis of the city. But the city is not a computer (Mattern, 2021); it is perhaps rather a 'machinic assemblage' in which 'the technical is not seen as separate from the social or the natural' (Amin & Thrift, 2002, p. 78).

Especially when approached from an atmospheric perspective, cities do not appear as partly material, partly technical, partly natural, and so on. The cities we explore here are everything at once; parks are both habitats for plant and animal life and essential places for human socialising, and streets are spaces of both transit and dwelling. There are elements that may momentarily stand out – a honking car, smelly dog excrement, or spectacular lighting – but the material and technical come together to make for environments that appear to people as whole environments rather than overlain aspects of the urban. In their wholeness, these environments also have affective qualities whose workings are formed by whatever is present in them at any given time. With that in mind, aside from their technical qualities, contemporary technologies like Liquid Light hold atmospheric power and thus have the potential to shape atmospheres and attune people to their surroundings – even if these surroundings go unnoticed or reside at the margins of attention. At the same time, cities are of course much more than the technologies that have become an increasingly noticeable part of their physical and affective makeup in recent decades.

This chapter is about the atmospheric qualities of urban environments as made up of a range of material elements, from solid forms like architecture, benches, and pavements to more elusive materialities like light, sound, and the weather. In it, we address how people are affectively entangled in atmospheres engendered by and embedded within multifaceted environments. These felt qualities of cities are shaped by technologies and urban design but also a variety of other phenomena, not least those elements of cities that are not human-made, but which nonetheless influence how spaces are sensed, for instance, presences usually registered in one form or another as 'nature'.

Certain elements in the overall urban environment are arguably needed for city life to function more or less smoothly, not least livable buildings and basic infrastructure. Other elements make life easier or more enjoyable: Benches, sports fields, urban green spaces for leisure, etc. And some elements are mainly in place to play the role of measuring our surroundings to inform urban policies, for instance, traffic counters and air pollution measurement tools. But none of these elements is ever entirely neutral; they are all wrapped up in urban politics, just as they all play a role in how places come to be constituted as *felt* to the people that dwell or move in them. In that regard, while much other research has scrutinised data-fixated and algorithmic notions of the city – including the many cases of flawed data (Mattern, 2021; Powell, 2021) – we attend to another issue here, namely how atmospheres in cities shape and are shaped by things, technologies, as well as more natural or ephemeral elements. This is in part in response to the shift of one of the central questions about architecture, and the city more broadly, from 'what does it look like?' towards 'how does it feel?' (Catucci & De Matteis, 2021, p. 16). That is, moving from a preoccupation with the visual sense of *looking at* to a bodily sense of not just *being in* the city, but of *being with* the city as attuned beings (Hasse, 2019).

We begin by entering the urban design studio SLA to get an idea of how they shape urban atmospheres through the meticulous selection of material components and surfaces. This leads us to explore how material things harbour atmospheric qualities that connect the immediacy of the present with the materiality of the past in urban settings. While such material encounters may be heavily designed by architectural companies, they are also affected by natural conditions like season and weather, and it is precisely a discussion of these conditions with which we end the chapter.

Atmospheric design

The Danish urban design studio SLA, named after its founder Stig L. Anderson, is renowned for its emphasis on 'city nature' – that is, not 'nature' as such, but a separate category of design intended to improve life within cities, including the life of non-human species. Design-wise, this often entails the appearance of 'wild' nature characterised by trees and plants not being neatly tamed or planted in symmetrical, straight lines (Stenslund, 2023). This is a central component of the company's fundamental emphasis on the *felt* qualities in their design: What they term '*det mærkbare*' (the felt) as a counterpart to the '*det målbare*' (the measurable). This is a company that explicitly aims to create aesthetic experiences through atmospheres (Andersson, 2014). Moreover, their work seeks to emphasise that humans are intrinsically part of nature. As they note,

> The quality of city nature does not depend on what it *looks like* [...] It is when we hear the birds singing, feel the decay of the trees, smell the fresh air following torrential rain, feel the storm, suddenly see the stars in the sky and physically sense the transience of everything, that we possess the most significant potential of feeling as one with nature.[3]

SLA's focus on the felt qualities of space became the starting point for our colleague Anette Stenslund's nine months of fieldwork in the studio, which we draw on here (see Stenslund, 2021, 2023; Stenslund & Bille, 2021). She shows how atmospheres have become central to the design process in the studio, from how collages and mood boards for internal use shape an urban design project, to how CGI renderings are moulded to make external stakeholders imagine the future (see also Degen & Rose, 2022; Degen et al., 2017). Through this process, before an urban space is even planned in concrete detail, let alone constructed, atmospheres guide visions and visualisations of future designs and thus play a crucial part in the design process.

To illustrate how this materialises in urban design, Stenslund (2023) explores the renewal of an urban space in Copenhagen that connects a shopping mall, metro, schools, and a library. In 2001, SLA was tasked with creating a more liveable urban space in this location, and to achieve this they made use of water, light, and sound design. In particular, the design of the soundscape can be seen as illustrative of the varied work the company has carried out over the years to invoke a particular 'city-nature' atmosphere and to spark curiosity among urban dwellers. Stenslund (2023, p. 112–114) relates how Anderson tells an anecdote about a mother and a little girl walking through the space when the girl hears a croaking frog, instantly invoking her interest in exploring where the sound came from. Following from that story, the space had seemingly transformed from a space of transit into a more playful space that encouraged people to dwell and to sense. To Anderson,

Stenslund (2023) notes, urban design is not (just) about what people do, but about shaping their sensuous and bodily experiences. As he comments,

> Whether or not the experience is awakened by natural or artificial frogs, it is real. Experiences can't be fake. The girl's nature experience is not any less authentic or intense just because the sound of the frog is staged and not caused by a real frog. It's just different. Neither greater nor lesser.
> (Stenslund, 2023, p. 113)

This particular urban space is not designed to take the shape of a forest or bog that people purposefully visit as such. But it is designed to affect people's moods, spark their curiosity, and spur them on to engage with the space differently from what they might have done otherwise. It may be that adults, all too familiar with urban spaces, continue to shut themselves off from their surroundings simply to pass through the space to get somewhere. But the little girl in the anecdote is vigilantly taking up the invitation to find the frog, in the same way she might have had it been an actual forest or bog as we know them. While the example does not tell of the disappointment that may have been inflicted had the girl found the loudspeaker behind the croaking, the point, to Stenslund, is that the atmosphere, 'derives its force for drawing in passers-by through elements of uncertainty and unpredictability as to what, where, and why the croak of a frog is present in a city centre' (2023, p. 113). As we will illustrate later in the chapter with the case of North West Park, another SLA-designed urban space, the main point is that urban design companies such as SLA treat atmospheres as an explicit concern, and aside from particular material surfaces and shapes, technologies like sound and light are key performers in shaping designed atmospheric experiences like the one described here.

From a broader perspective and in line with Stenslund's work, Mark Wigley has noted how atmospheres hold a central place for architecture and writes: 'Atmospheric design is itself the product of a particular atmosphere' (1998, p. 26). Similarly, and in line with our perspective, rather than simply understanding 'architecture' as the sum of its tangible parts, when people experience architecture, '[w]hat is experienced is the atmosphere, not the objects as such' (Wigley, 1998, p. 18). The wider implication of this argument is that when stepping into a room, a person will not necessarily primarily *see* a dimly lit bed, *smell* a candle, and *hear* the soft music playing, but rather *feel* an atmospheric space conjured up via those things. Whether that atmosphere is then evaluated as pleasant and romantic, exciting, threatening, or gives cause for anxiety is a matter of context and subjectivities. The presence of an atmosphere and how one relates to it are two distinct issues. We are here reminded of Böhme's (1993, p. 125) phenomenological starting point:

> The primary 'object' of perception is atmospheres. What is first and foremost perceived is neither sensations nor shapes or objects or their constellations, as Gestalt psychology thought, but atmospheres, against

whose background the analytic regard distinguishes such things as objects, forms, colours, etc.

Importantly, to Böhme, these atmospheres can be staged. The production of atmospheric spaces and their experiential results are not by any means objectively given in advance but are nonetheless entangled in the architects' design traditions and notions of how whatever is being designed *should* feel. Similar forms of atmospheric staging are also carried out by others far beyond the architectural field. While architects might undertake this sort of task when choosing building materials, interior designers do it when refurbishing a building, restaurant workers do it when clearing the table, putting everything in order, replacing the candlelight, and 'resetting' the scene for a new quest. And people do it to a greater or lesser degree when they get up in the morning to set the table, turn on the radio, and pour a cup of coffee before commencing breakfast. Once this atmospheric work (cf. Ahmed, 2014) – or type of deliberate 'atmospheric practice' (Bille & Simonsen, 2021) – has been made, the stage is set.[4]

Ultimately, atmospheric cities are experiential phenomena that are being shaped in large part through design. While attunement is an ongoing process, and atmospheres are porous and ever-changing, designed elements make up a significant part of the material foundation on which cities come into being. At the same time, as pointed out above, the relationship between atmosphere and design does not only play out in urban spaces post-construction. It is increasingly becoming built into design processes as architects and others start their work from notions of what a future space *should feel like* and thus let loosely shaped ideas of future atmospheres guide their projects. Yet, while atmospheres may be staged by design, 'what they are, their character, must always be felt' (Böhme, 1998, p. 114).

Feeling the atmospheres of urban design

Intentionally trying to produce atmospheric impressions may not result in spectacular design interventions such as installing neon signs in Times Square or at Piccadilly Circus. Atmosphere-oriented design is sometimes reflected in forms much more toned down and less tantalising. For instance, in central Stockholm, the refurbishment of the square Brunkebergstorg right in the heart of the city is an illuminating example of a contemporary redesign project that centres on an atmosphere-heavy aesthetic expression. This square was redesigned in 2017 by the firm Nivå for Stockholm Municipality. Unlike many other places we have visited in this book, very few people live around Brunkebergstorg. Rather, this part of Stockholm is characterised by the imposing presence of stores, shopping centres, and office buildings. At the square itself, a few restaurants and several upscale hotels have been thrown into the mix and seem to underscore that this is much more than a residential neighbourhood.

Figure 3.3 Brunkebergstorg, Stockholm.
Photo by Siri Schwabe.

The entire block facing the square toward the east is being turned into an upscale neighbourhood hub, which has been narrated and branded as an 'urban escape'; 'a space where people want to be. Where everyday meets extraordinary and where workday meets leisure time in a brand-new way.'[5] The square itself features a number of brass light fixtures and fountains in perforated metal that light up after dark, with the light emitted being soft and evenly distributed. When we first visited in December 2018, a Christmas tree as well as festive lights had also been temporarily set up in the shape of moose and deer, attracting a number of visitors who came by to have their photos taken in front of them.

This design marks a big change from what the square was before. Brunkebergstorg has a history as the site of both seedy and subversive activities, from being the starting point of one of Sweden's most bloody uprisings in the mid-1800s to being the prime location for illicit business in the twentieth century. At the same time, up until the 1950s, the square was something of a meeting place for the political and cultural elite. It was also an important node in the transport infrastructure of the city. For many years the square acted as a main thoroughfare, and it was previously a main hub for city buses starting in the 1930s. Brunkebergstorg has always held a central position in the city in more than one sense, then. In the latter half of the twentieth century, however, Brunkebergstorg ceased to be the gathering point it once was. Beginning in the 1950s, much of central Stockholm changed as it was redesigned according to modernist, functionalist ideals, which did not leave much space for the old built environment. And in the early 1970s, as part of this near-complete remodelling of the city centre, the massive culture house simply named *Kulturhuset* was inaugurated right next to Brunkebergstorg, effectively creating a massive barrier and leaving the square hidden away at the back of this new heart of the city, leading to significant consequences for the square and the way it was being used.

One of the people we spoke to at the square had worked in the area for a number of years in the late 1970s and early 1980s when both Brunkebergstorg and its surroundings looked very different. Back then, according to him, this was not a place anyone would go to unless they really had to. The general perception back then, he explained, was that the area was characterised by the presence of drugs and prostitution, and was a 'rough' and 'dangerous' area. Now the sex work, the begging, and the drug trade seem to have essentially vanished from the area, creating something much livelier and more welcoming out of what, to many, had been perceived as a 'forgotten' and 'dead' place.

The history of Brunkebergstorg unsurprisingly seemed to affect the perception of the place by those we talked to who knew about it. However, even the people we interviewed on the square who had little or no familiarity with what it used to look like described the new design as 'clean', 'neat', and 'open'. Although their opinions on the square in general differed – most people liked it, some did not – most agreed that the square had been 'opened up' by the redesign and made 'lighter'. According to one older woman we spoke to in a voxpop, despite the square being surrounded by largely tall and dark buildings, being able to see the sky above and between the buildings lent the place an airy feeling that it did not have before when the greenery was denser. The general consensus among those who knew the place before seemed to be that the square now felt safer and more welcoming. In the words of one person, it had become a place that is 'open to everyone' – although, we might add, with the recent exception of course of those drug dealers and sex workers that used to do business there.

Ultimately, the redesign of Brunkebergstorg has been intended to open the space up perceptually and effectively does so to our participants. It has more or less eliminated dark corners and thus exposes what takes place on the square to its surroundings. It does so not by spreading light evenly across the square, but by a careful orchestration of lighting and an overall design that carries atmospheric potential. Although the layout of the fountains and other material elements present on the square denies visitors the possibility of getting a full overview of the space from any one particular location within it, the absence of trees and other similarly tall and imposing features on the square itself means that a feel of openness is nonetheless prevalent among its users. In this case, designing for atmosphere has been a step in the process of creating a space that is both more physically and socially open than it was before.

The example of Brunkebergstorg helps us think about the role of visibility in shaping experiences in urban spaces. Writing on Potsdamer Platz in Berlin, John Allen (2006) explores the design of a commercialised public space where the atmospheric effect is emphasised. He argues that a novel kind of urban space is emerging where power works 'not through electronic surveillance technologies or some rule-bound logic imposed from above, but through the experience of the space itself, through its ambient qualities' (Allen, 2006,

p. 442). 'Ambient power', he notes, taints the experience of spaces and 'seeks to induce certain stances which we might otherwise have chosen not to adopt' (Allen, 2006, p. 445). Allen explores the way a privatised – albeit seemingly public – urban space may appear as democratic and open, attending to a range of needs, while working through seduction to make people interact and circulate as well as to limit unwanted behaviour. This type of seductive space, he suggests, showcases a kind of 'ambient power' that is becoming increasingly prevalent and that functions in more subtle ways than what we might associate with more traditional forms of urban power dynamics as they play out through, for instance, overt surveillance strategies, including via lighting (Schivelbusch, 1988). The use of ambient design features for seduction has been made possible not least with the advent of smart lighting technologies and the possibilities of inserting colourful light diodes almost anywhere, enveloping the most mundane parts of cities in an aesthetic nocturnal wrap. This shift in urban lighting has been conceptualised through the distinction between the so-called disciplinary and the spectacular paradigm, the latter of which highlights how 'illumination is seductive and dazzling, creating the stage on which the commodity makes its breathtaking appearance: light is deceptive and narcotic' (Otter, 2008, p. 2).

Following from this approach, the promise of visibility – and the related potential of surveillance, with all the disciplinary effects that entails – comes to work at Brunkebergstorg and elsewhere as a designed seductive measure evincing an ambient power. This is not a case of blatantly spreading light to allow for visibility, but more about scenographically designing the *sensation* of an 'open' space. With that, Brunkebergstorg is an example of how urban design, and lighting design in particular, is being used to aestheticise and create particular atmospheres in public spaces, making these spaces appear welcoming and attractive while harbouring a subtle type of power. Much like how prominent buildings are increasingly being externally lit to showcase them in the dark hours, so too are urban spaces more and more often lit for aesthetic, sensory, and atmospheric effect.

At the same time, the atmosphere thus nurtured relies on particular presences, many of which are material; made up of bricks and mortar, stone, concrete, and metal. It is to these designed *things* and their affective qualities we now turn.

The affective qualities of things

The atmospheric qualities of things have been the focus of much recent research in the social sciences. From looking at the everyday life among ruins in Cyprus (Navaro-Yashin, 2012) to affective infrastructure in Nigeria (Larkin, 2013) and a 'Nordic bias for what makes spaces comforting' (Catucci & De Matteis, 2021, p. 15), the ubiquitous presence of buildings, tangible technologies, and urban design features throughout contemporary cityscapes must be taken into account if we are to move toward an understanding of how atmospheres and urban environments work in tandem. Buildings most often host both human and non-human life, reflect and manage weather conditions, and have a distinct effect on how public urban spaces are shaped in that they mark the negative space around them. As things, they are more than simply material; they are also materialising social and atmospheric qualities whose impact on lived cities is difficult to overestimate (see Gieryn, 2002). Meanwhile, although buildings should thus not be seen as somehow separate or blocked off from the surrounding city – indeed they are permeable things – they do in most cases mark the border zones between spheres such as private and public.

If we move our gaze slightly away from these unavoidable giants of the urban environment and toward the spaces around them, sites such as streets, public squares, and parks often feature a wide range of material objects and infrastructure. They most often include some form of seating in the form of benches or other elements of a similar height. They might have areas dedicated to lawns, trees, or flowers. Some have been constructed around public or semi-public institutions, train stations, or similar. Others might feature a centrepiece in the form of an ornamental fountain, a statue, or a sculpture. Further, plazas and parks practically always include some kind of fixtures emitting artificial light that serves to secure visibility after dark and that likewise, as we saw above, affects how these places feel.

To give a perhaps extreme example of the latter, in 2013, the Oslo neighbourhood of Stovner saw the addition of an oversized lamp to a large green area adjacent to a number of tall, grey, concrete housing blocks. Since adopted into the Guinness Book of World Records as the world's tallest floor lamp,[6] it measures 9 m and 16 cm in height and features a base that doubles as seating. In contrast to its material surroundings, which appear somewhat drab, the lamp is bright red with an orange polyester and fibreglass shade that gives the light emitted from the installation a warm glow.

Figure 3.4 The oversized lamp in Stovner, Oslo.
Photo by Oda Fagerland.

During a stroll around Stovner with Maren, a woman in her late 20s who had grown up in the area, she described the lamp as reminiscent of something that would go in a country-side cabin. It had a distinctly 'grandmotherly' shape to it, she said, as if it – or rather a much smaller version of it – belonged in an older person's quiet home rather than in the middle of an urban housing area. Her relationship with the neighbourhood was an ambivalent one; she was both nostalgic about her upbringing in the area and quite happy to have left it to live elsewhere. In her view, installing a massive lamp seemed a meagre attempt to improve the feel of the neighbourhood, noting that perhaps it would have been more helpful to invest in local schools and other institutions instead. As she reflected, art installations like this one are 'of course never popular because the money could have been used elsewhere... When you don't get funds for other stuff it's tiresome that [money] is spent on that, that goes without saying.'

At the same time, as darkness was falling during the walk, Maren declined to take a seat on the nearby benches facing the lamp, and opted instead to go into the light to sit directly beneath it, noting how it felt simultaneously a bit sad and a bit cosy and remarking that 'it's probably because we're sitting here on the bench underneath such a warm light'. Despite her thinking the money that had gone into it might have been spent better, Maren's experience of the area as well as her movements within it were nonetheless shaped by its inevitable presence. She found herself being pulled toward its warm light in an otherwise dark environment after dusk and characterised it as adding a cosy homeliness to a spot that did not have much of that previously. Not only had this lamp become an imposing feature in the urban landscape, but it also worked as an atmospheric feature, and with that, affected how someone like Maren became attuned to her surroundings. In other words, it changed her way of being present in the space along with its transformation of the space itself.

To a very different effect, in North West Park in Copenhagen, a bench has been placed directly under a white spotlight. The result is scenographic, yet both the perceptual and felt qualities of actually sitting there, in all the glare of the light, is less than comfortable; the bright light simultaneously blinds anyone seated there as well as exposes them to their surroundings, which are obscured by darkness in contrast.

During the hours of fieldwork spent in the park, we never observed anyone sitting on the bench after dark. While the bench in North West Park itself is recognisable with its iconic design that is used widely across the city of Copenhagen, the bright spotlight allows it to play a much more dramatic aesthetic role during the dark hours by shaping the feel of taking a seat on it. Put somewhat differently, it is in the relation between the thing – the bench – and the immaterial presence of light that a particular atmosphere arises, at particular times. Even if not intended, this specific place in North West Park seems to be essentially barring the experience of comfort, leaving potential users so uncomfortable – physically, but also emotionally and socially – that they opt to simply avoid this particular place in the park, at least after dark.

Embraced by the city 63

Figure 3.5 Illuminated bench in North West Park, Copenhagen.
Photo by Mikkel Bille.

Instead, as we observed during our fieldwork conducted across all four seasons, other parts of the intricate park become more attractive to those spending time there, while the majority of its users simply move through the space rather than dwell there for long after darkness falls.

A very different example of the atmospheric potency of something as mundane as a bench is the highly praised and award-winning 'Camden bench', which was installed in certain parts of London in 2012. To the naked eye, the Camden bench resembles little more than a slab of concrete, its rectangular shape somewhat reminiscent of a coffin, albeit with a surface for seating that slopes down at various angles rather than appearing uniform and flat. However, the subtleties of its design have been deliberately put in place for a reason. Designed to deter unwanted presences and activities, the bench is really just that: A bench meant for seating and entirely unsuitable for anything else. The surface of the material repels graffiti and other types of vandalism, the impermeable flatness of its top and sides leaves no secret spaces to hide drugs or other such unwelcome substances, and its angular shape is meant to put off skateboarders from engaging it as part of their urban stage. Perhaps most significantly, the sloping top makes this bench an uncomfortable choice for those – homeless and other rough sleepers – seeking rest.

While the illuminated bench at North West Park is perhaps a somewhat ambiguous case of how design can be simultaneously accommodating and dissuasive to use, the Camden bench is a poignant example of what has been called 'hostile' (see Rosenberger, 2020), 'unpleasant' (Savic & Savicic, 2012) or 'dark' design (Jensen, 2018, 2022): Design that highlights the exclusionary potential of the distinctive material forms that take up urban spaces. Somewhat paradoxically, the removal of unwanted human presences by designing for discomfort is tied to notions of what makes spaces comfortable for the *right* people; those who use spaces in the *right* way and whose presence is not understood to produce unease in others. In other words, it seems the idea is that angular benches and other types of dark design are to be preferred over the potential discomfort caused by 'undesirable' users and their ways of occupying a shared space. Indeed, this type of design seems to be about creating comfort zones in which comfort is exclusive. In that regard, it is important to remember that 'comfort can be weaponized' and that discomfort can be generative of change as it alerts people to the challenges of coming together across differences in urban space (Owen et al., 2022, p. 1). Sometimes, this weaponisation comes in the shape of physical objects designed to deter some uses, while other times, as described further above in this chapter, it works more subtly and seductively as a kind of ambient power (Allen, 2006) through, for instance, the scenographic staging of lights.

Designs such as those brought up here make it abundantly clear that in experience, human bodies and their environments are difficult to separate. Bright lights that put people off dwelling in their glare and benches that only afford a certain type of bodily engagement – that of sitting upright – underscore how the material qualities of urban spaces shape people's ways of being in them; their uses of and practices in those spaces, of course, but also their affective engagement with them – the broader sense of being in a space where the design explicitly deters particular kinds of behaviour. In other words, designed things at various scales – from buildings to light fixtures and urban furniture – play a crucial role in the atmospheric city as they facilitate both how cities are used and how they feel. Moreover, they do so in a differentiated way whereby urban spaces come to be perceived as welcoming and unwelcoming, accommodating and exclusionary to different people at different times, not just as material spaces but as atmospheric spaces.

Sometimes these interplays between material things and affective experience are starkly noticeable, as they might be to a homeless person looking for repose and being faced with an atmosphere of rejection through the Camden bench (Jensen, 2020). Often, however, people move through cities without reflecting on how the physical environment around them might affect their feelings and perceptions in and of space. Be that as it may, material elements are urban presences that evince the atmospheric qualities of things and that at times seductively and subtly shape the power dynamics of the atmospheric city. At the same time, without such things and infrastructure, cities would not be cities. As we explore further below, these material elements are also part of the stuff of urban spheres that grounds cities in time.

Environments of atmospheres past

In describing the atmospheric qualities of a seventeenth-century row house district – an old and repurposed army barracks – in central Copenhagen, the author Amalie Laulund Trudsø (2013, p. 9, our translation) notes that: 'You are thrown back by Nyboder, or: On a walk you are carried forward by history, you are pushed ahead by the atmosphere instilled in you by the yellow walls.' Nyboder stands out among its architectural neighbours by being made up of extensive rows of relatively low, two-storey constructions all covered in the warm yellow colour that simultaneously binds them together visually and sets them entirely apart from the taller and more diverse apartment buildings that surround them. In this part of the city, passersby are confronted with the past as a striking feature of the built environment that, at least in cases like the one described by Trudsø, affectively pushes them further into the atmosphere of the neighbourhood.

In a wider sense, through both designed and natural elements, the presence of the past plays a significant role in shaping the atmosphere in cities. Aldo Rossi has written that 'the union between the past and the future exists in the very idea of the city that it flows through, in the same way that memory flows through the life of a person' (1982, p. 130; see Hebbert, 2005). Similarly, in his much earlier work on collective memory, Maurice Halbwachs (1980; first published in 1925) highlighted the importance of space to how people perceive and understand a shared past. Places, to Halbwachs, are imbued with collective memory and, with that, identity. In this perspective, perceptions of the past are shared in common experiences of places, and, at the same time, the past itself is understood through images and concepts of space.[7] In his view, the past is in fact imprinted on space (Halbwachs, 1980; Hebbert, 2005, p. 584). This imprint, it could be argued, is largely an atmospheric one, as the example of Nyboder above intimates. Indeed, the material and immaterial heritage that surrounds residents and visitors to urban centres raises questions of how they attune and become attuned to spaces that have different histories and potential futures, as well as how atmospheres and particular histories reflected in the urban environment interrelate.

Stockholm Royal Seaport and its surrounding harbourfront areas are perhaps a case in point. At this extensive location on the outskirts of the city, the old, the new, the dilapidated, and the halfway-built intermingle in the landscape. While many of the locals are preoccupied with the tensions that arise when new residential areas are added to an existing enclave of housing as is the case here, some people are more concerned with the future of run-down industrial-era features.

To Björn, for instance, plans to renew the expansive port well beyond the existing residential neighbourhood made him worry that the atmosphere of the largely abandoned and decaying environment of the old harbourfront would be lost. Björn, a young man in his mid-20s who had been living in student housing overlooking Stockholm Royal Seaport for several years by the time we spoke to him, described how he enjoyed taking solitary walks around the old port area in the dark, soaking up what he called the 'apocalyptic atmosphere' down there. Being an art student, Björn took inspiration from the worn and partly ruined buildings and yards along the water and had formed a habit of collecting bits of debris from the area to incorporate into his art. In his view, the sort of atmosphere he experienced there should be safeguarded. In other words, he believed the built environment would be better off left alone to wither with time.

Every city has spaces that were once occupied and have since been abandoned, whether for shorter or longer periods of time. Decay and ruination, although often indicators of human absence, do not necessarily make for spaces emptied of meaning, however. Often, as in Björn's case, it is quite the contrary. To Björn, it was the decaying material heritage of the old port that allowed him to experience the area in a way that resonated with him and which he feared would be lost if these elements of the built environment were to be replaced with something newer and presumably neater. In other words, the ways in which the past was manifested in the environment of the harbourfront played a significant role in shaping an atmosphere that Björn, not unlike the people we met in Chapter 2, intentionally sought out. That ruins and left-behind structures of the built environment harbour certain atmospheres and carry special qualities is not specific to this particular case (see Edensor, 2005 and, for broader perspectives, DeSilvey, 2017; Mah, 2012; Stoler, 2013) but the extensive nature of the Stockholm Royal Seaport development and the complexities of how its past is dealt with offers depth to our understanding of the entanglements between urban history, the environment, and how cities feel.

Meanwhile, for many of the neighbourhood's residents, the past was present to them in a different way, despite the ongoing and enormous changes to their surroundings. Indeed, we spoke to several people who described their connection with Stockholm Royal Seaport as being tinted by their memories of childhood. In those cases, people tended to bring with them certain embodied memories of what the area used to feel like when they would visit the nearby forest-like park or the old workers' village on family outings as children. Older residents in particular recalled winters of skiing in the park, taking in the seemingly conflicting smells of the natural environment and the now out-of-use gasworks that characterised the old industrial compound of the neighbourhood. Engaging with the current-day atmosphere of the area in that sense entailed engaging with what one of us has elsewhere called *atmospheric memory* (see Schwabe, 2021), whereby the present-day feel of the neighbourhood came wrapped up in memories of past atmospheres.

To many of the area's new residents, it was indeed their past connections with the area and the memories associated with it that motivated them to move there when new apartment buildings started springing up around the gasworks in an extension of the old workers' housing. Even those of our participants who did not have an established connection with the neighbourhood sought a sense of attachment by engaging with its past, reading up on its industrial-era history and looking into the role of the park – which had previously functioned as a royal hunting ground – in the development of the neighbourhood and wider city. Engaging imaginatively with this past, as they described to us, had motivated them to take active part in developing the neighbourhood as a social space and to begin to build a sense of community among the many newcomers, whether through the establishment of residents' associations or enthusiastic support for local shops and other businesses. They were taking active part in continuing the process of developing ruins into something else, changing the remnants of the past – material and atmospheric – in more or less subtle ways along the way.

Figure 3.6 Stockholm Royal Seaport.
Photo by Siri Schwabe.

Taken together, these examples highlight how manifestations of the past in urban environments affect present-day atmospheres in two ways, namely through heritage – including architectural heritage, both ruined and maintained, mundane and monumental – and through memories as well as other forms of imaginative engagement. Indeed, both a shared urban heritage, personal memories, and imaginative investment play a crucial role in this sense of pastness. As regards people's relationship with the built environment, time also plays a significant role in a number of other ways. Besides people's own moving through the temporal world, ageing and changing over time and thus constantly revising how they relate to their surroundings, the grounding of the built environment in time is also a significant factor in terms of how it is experienced in day-to-day life. Whether a building, an urban space, or an entire neighbourhood is deeply settled or newly built, undergoing a process of decay or being still under construction, its place in time affects how it comes into being as part of an atmospheric city. This means that not only its visual qualities are emblematic of a style or design, but its atmospheric qualities establish the past and presentness of the city (cf. Hasse, 2010). In this sense, the sense of pastness and presentness also becomes embedded in processes of decay, renewal and maintenance.

Weathering the atmospheric city

So far, our focus in this chapter has largely rested on the materiality of cities and how designed elements, whether new, maintained, or dilapidated, interplay with atmospheres to make for urban environments that are felt, affectively embracing people in often subtle ways. In this final section, we shift our focus to the ever-changing phenomenon of the weather and its role in shaping atmospheric cities, from rain and sunshine to the eternal presence of air, whether in the form of a fresh breeze or a haze of stale dust or pollution. We do so because the atmospheric city is continuously being constituted in the coming together of a plurality of elements – material and immaterial – that may be separated conceptually but that overlap in everyday experiences. The weather as one such material, yet not necessarily tangible, phenomenon is always mediating such experience. We are always immersed in the weather – in what Tim Ingold calls a weather-world (2011). In other words, cities are not just the designs, technologies, materialities, and people that fill their spaces; they are also a host of non-human and intangible presences, not least those associated with the term weather.

At the same time, weather also plays an increasingly central role in urban development projects and has been a driving factor in regard to design interventions at various scales that highlight the necessity of mitigating the effects of climate change and extreme weather events. In Copenhagen, there is a current cloudburst protection policy that aims at improving the urban infrastructure of *c.* 300 squares and areas around the city to handle the rise of downpours and storms, with billions of Danish kroner in investment. At the time of writing, several large-scale urban regeneration projects are in progress in the city, and the ongoing development of a new human-made peninsula in extension of the Copenhagen harbour – known as Lynetteholm – has been framed in part as a response to the need for securing the urban coast against rising sea levels, although this discourse has also been heavily criticised. With that, it is not only in experience, but in concrete design and planning practice that the weather makes the city.

We can already see how weather and weather-related events affect people's ways of being in and feeling their urban surroundings. In Copenhagen, Jens, a retired consultant in his 60s, fondly described his experience of getting caught in a thunderstorm while out and having to seek shelter in a doorway alongside strangers in the same situation. The weather, in a very concrete way, spurred Jens on to occupy the city differently, at least temporarily, and it facilitated a different kind of encounter for him, both with strangers and with the urban environment in a broad sense. It also allowed him to feel the city differently. Reflecting on a more mundane phenomenon, namely cycling through the city, Jens told us how the weather influences his experience:

> I mean, if it rains, I won't take my bike… I think of course the weather matters to what experience comes out of it… because sometimes the bike is just a means of transportation from A to B. And if it is bad enough, the important thing is to move fast. Then, you don't really see all that much. But if the weather is nice and it's warm, it opens you up to impressions to a greater degree.

In both instances, rain is the stand-out weather phenomenon, and the one that most blatantly transforms otherwise more ordinary ways of being in the city. People are immersed in the weather, and sense the world around them through and not just in the weather (Böhme, 2011; Ingold, 2007). In Denmark, it rains an average of 179 days per year, and experiencing the city as damp and wet is thus a common occurrence. At the same time, we see with Jens how the advent of a rain shower or downpour makes all the difference to him, at least for a time. It affects his movements, what and who he comes into contact with, and it influences what and how he perceives the city and his own way of being in it – a weather-world, 'in which all are immersed, and in which nothing ever stands still' (Ingold, 2007, p. 34). In other words, it not only impacts, but indeed is an integral part of his physical, sensory, and emotional presence in the city. With rain, the urban environment looks different (imagine looking at a street through the haze of heavy rain), sounds different (imagine the sound of car tyres on a wet surface), smells different (imagine the smell of petrichor as the rain hits the warm asphalt in summer), and the tactual sensation feels different (imagine the slippery surface of the pavement as you turn a corner on the bicycle).

While precipitation had, in Jens' experience, had an immediately felt effect on his way of being in the city, he described to us how changing light conditions tended to work in subtler ways. From the home he shared with his wife on the harbourfront at Nordhavnen, he enjoyed a view of the sea and could see the Swedish port city of Malmö across the water on most days. However, as he explained as we spoke to him one day around dusk:

> The light changing means that the view changes constantly. This time of day, when the sun sets, we see the lights blinking over in Malmö when the sunbeams hit [the landmark building] the Turning Torso. That's what

light does; it highlights something that you wouldn't otherwise notice or see. The angles change, the shadows change.

Jens continued to reflect on light more generally, describing how, in the 'low light' provided by the sun just before it disappears below the horizon, 'everything becomes a bit more than just three-dimensional'. Whether in the woods or in the city, he said, 'you feel that there's a certain depth' to what is being observed, giving it 'that effect – the more-than-three-dimensional'. Jens' description is evocative in its suggestion that certain types of light lend depth and further substance to the environment as it is perceived. It also points to the entanglement between physical landscapes, ephemeral phenomena such as sunlight, and the embodied attunement processes that place people experientially within their surroundings. In this phenomenological perspective, the weather 'is a whole that corresponds with our corporeal-sensory feelings' (Frølund, 2018, p. 159). It is a bodily experience and immersion that works through the senses as it is smelled, touched, heard, and seen, and that, as a totality, wraps itself around people and the city.

Another resident of the harbourfront in Copenhagen noted how the water gave a 'dynamic' vista of the city that offered a sense of openness. On sunny days it facilitated a clear view and on other days there was a cloudy haze covering the top of the spires and towers of the city. 'It has a dramaturgical feel in some way.' She summarised that 'it's important to me, to feel at home, that it's open', only to heavily critique other areas of Copenhagen such as the OMA-designed Bloxhub building and the central pedestrian high street close to the city hall with its many tourists, small shops, intense city life, and secluded spaces: 'I couldn't imagine a more horrendous space.' The weather and the dynamic horizon allowed her to feel the city as open, thus resembling the experience of the people at Brunkebergstorg that we described earlier.

Now, there are of course a plethora of other weather conditions than simply rain and a visible sun around dusk. The examples above did not include the role of temperature, which also makes a marked difference to experiences of precipitation as it takes the forms of an all-day light rain in the fall, a brief summer downpour, or winter snow or sleet. However, the key point is that the weather has an atmospheric effect on the city and the people who inhabit it; people who may be attuned to spaces designed to feel in a particular way, but which are also rather unpredictably co-defined by other elements, not least as defined by the weather.

In short, the weather is an inherent part of the city and people's way of being in them. Indeed, experiences of the urban environment are shaped by the ways in which the weather impacts the very materiality of the city by tearing on surfaces, corroding metal, flooding streets, and blowing garbage around. Yet more fundamentally, the weather is an elemental part of what makes the atmospheric city porous as the wind, sun, and rain penetrate material and human bodies. As Tim Ingold notes, 'to inhabit the open is to dwell within a weather-world in which every being is destined to combine wind, rain, sunshine, and earth in the continuation of its own existence' (2007, p. 20).

Conclusion

> There are these two young fish swimming along and they happen to meet an older fish swimming the other way, who nods at them and says 'Morning, boys. How's the water?' And the two young fish swim on for a bit, and then eventually one of them looks over at the other and goes 'What the hell is water?'

So David Foster Wallace (2009) illustrates that 'the most obvious, important realities are often the ones that are hardest to see and talk about'. The air people breathe, the ground they stand on, the buildings around them, the clothes that wrap their bodies, and the light and sound that embrace them all come to situate them in place, whether or not in ways that are agreeable. Most of such material phenomena will often go unnoticed, like the water to the fish. Some will stand out at times, like the aroma from a bakery enveloping everything in the vicinity in a sweet smell. And some material phenomena offer resistance or challenge people's embodied presence, such as slippery ice on the pavement on a cold winter morning. But most things are in all their embracing capacity humbly located at the margins of attention (cf. Miller, 1987).

In Nordic cities and beyond, urban dwellers find themselves embraced by an environment that encompasses phenomena such as weather, plant and animal life, as well as varying materialities: The tarmac of city streets, the ever-present structures of concrete that mark urban spaces, and the numerous other designed and human-made elements that make up the built environment. In addition, as we have discussed here, the materialities of the city play an immense role in shaping this overall environment; they inform and guide movement, they are everywhere present in the sensuous impressions they offer, the noise they emit, and the multitude of other output they produce around the clock.

The urban environment, then, is difficult to pin down in any straightforward way; it is a conglomeration of myriad elements in constant flux experienced by people who, with all their differences, undergo changes along with the city as well. Under such circumstances, urban environments often appear to people in simultaneously conspicuous and subtle ways; they require some attention if people are to navigate safely within them, yet the non-human and material presences that characterise them are most often so numerous and diverse that it becomes all but impossible to take them in other than in a subtle, peripheral manner. Indeed, an urban environment is much more than its notable elements. It is also a felt atmospheric environment. In a phenomenological sense, it is largely through atmosphere that people relate to urban environments and, conversely, these very environments play a significant role in constituting the atmospheric city overall.

Urban spaces, their materialities, and their designs, albeit never entirely fixed, are subject to slower change and generally show less dynamism than the attunement of people and their relations. Urban design and infrastructure most often evince a certain stability, or what we might, with Anique

Hommels (2005), call obduracy. In such relatively stable material environments, we might find a sort of spirit to a place – a *genius loci* (Norberg-Schulz, 1980) – playing an important role in how people make sense of its past, present, and potential future. Architecture, technologies, and design matter in shaping these environments. Yet, this sense of place is never truly fixed, as people move through urban spaces, occupy them, or avoid them. Also out of the hands of architects, planners, and even people, are the passing of time and the weather. The latter may on the one hand change rapidly, or it may be steady for several days as a meteorological condition. Yet, the weather and its impact over the course of a day – how the sun heats seating areas, how rain puddles muddy up otherwise green spaces, or how an evening wind picks up dry leaves shed by nearby trees in autumn – embrace people in weather-worlds through sensory attunements.

The point of this chapter has been to pinpoint how the atmospheric city is constituted by what makes up an urban environment beyond the human. This constitutive process largely centres on the seductive qualities of design and technologies that help shape the feel of a place. Yet, these designs and technologies do not alone determine how places feel. Indeed, atmospheres also depend on the passing of time and local weather conditions, not just because the night makes it more difficult to see or rain makes surfaces slippery, but because people do different things at different times, and because the weather taints perception through and through.

Notes

1 In 2019, after the redesign of Østbanehallen, ÅF merged with Pöyry to become AFRY.
2 https://afry.com/en/liquid-lightr. Accessed 19 November 2021.
3 https://www.sla.dk/perspectives/city-nature-its-how-it-feels-and-functions-not-how-it-looks/. Accessed 16 May 2022.
4 A note on Böhme's terminology: Stating that what people first and foremost experience is atmospheres, only to suggest that atmospheres can be produced, may seem an oddity. Years after the publication of his first book on atmospheres (Böhme, 1995), he makes a distinction between 'atmospheres' as felt and 'atmospheric' as that which is designed, stating that 'atmospheric phenomena are distinguished from the atmosphere by the substantial absence of subjective moments' (2001, p. 60, our translation).
5 https://www.amffastigheter.se/en/urban-escape/discover-more-of-the-neighbourhood/. Accessed 26 April 2022.
6 https://www.guinnessworldrecords.com/world-records/largest-floor-lamp. Accessed 2 May 2022.
7 See also Bakshi (2017) as well as Jones and Garde-Hansen (2012).

Figure 4.1

4 Moving through atmospheres
Mobility and attunement

'You constantly need to be aware of other people in the bike lane, because quite often there's like this anarchic atmosphere where people go super-fast, and you need to be attentive all the time and look ahead.' Kim, whom we met in Chapter 2, was generally comfortable biking in Copenhagen. It was part of everyday life, but also a form of mobility that entailed a particular mindset and atmosphere, requiring cyclists to constantly pay attention to the movement around them. Kim continued: 'If I sense that everyone else is in a kind of morning tempo and bike really fast, then I bike fast too.' This experience of moving through the city was markedly different from walking, as Kim explained: 'When I walk it's maybe more like I'm more quiet inside my head. I don't think so much about all sorts of things. In that way I can just walk and notice things in a way.' As we see with Kim, movement shapes the way people attune to places and allows for transitioning between different atmospheres of the city. At the same time, movement itself – that of ourselves as well as that of others – also shapes the feel of the city, whether this feel takes the form of a hectic morning atmosphere in the bicycle lane or the internal quiet fostered by a slow walk.

This chapter takes the starting point that atmospheric cities are cities in movement and investigates the particular role of movement in constituting them as felt environments. Our interest in movement reflects a broad understanding of the term as encompassing both the physical, observable movements of humans and non-humans through space via various forms of mobility, as well as the less easily discernible ways the world appears to people through atmospheres that are porous and fleeting. Atmospheres, in that perspective, unfold through their capacity to move people and to be felt by them, while people conversely also shape them through their movement and actions. In other words, although atmospheres rely on a range of design and non-human factors to take form as illustrated in Chapter 3, the movement of people and things charges atmospheres and affects people's practices and choices with regard to mobility. Public transport may, for instance, be rejected if it feels unsafe or cramped; alternative routes may be taken on foot if the direct route means passing an eerie industrial area; hectic bike lanes may spur you on to increase your own speed, or cars may be preferred for the intimate space they offer to just sit alone and listen to the radio while moving. The atmosphere matters.

DOI: 10.4324/9781003379188-4

The central argument in this chapter is, then, that how cities feel depends on how people and their surroundings move. While Kim did find a qualitative difference between walking and cycling, we cannot, based on our data, more generally say that one form of mobility necessarily entails any one particular atmospheric experience. Indeed, walking can be hectic and cycling might offer a space of tranquil introspection, all depending on the context, as we will show. With that, the point from an atmospheric perspective is not so much that a specific practice defines attunement across the board, but rather to show how particular processes of attunement unfold through movement. In that sense, our concern is not with contributing to the growing literature and journals on mobilities (cf. Bissell, 2018; Dennis & Urry, 2009; Sheller & Urry, 2006; Urry, 2012; Vannini, 2009), but with dwelling on how a view towards movement helps us understand atmospheric cities as lived (cf. Catucci & De Matteis, 2021; Gandy, 2017; Hasse, 2012).

Movement and mobility are fundamental aspects of city life and relate to deeply political issues regarding urban development and questions of how cities and their infrastructures should be constructed to deal with the human inclination to move. Moreover, getting around cities efficiently is consistently highlighted as a crucial component of living a good urban life. Walking, in particular, has become a central feature of recent ways of thinking about the city, such as Jeff Speck's *The Walkable City* (2013) or Carlos Moreno's *15-Minute City*: A city where everything one might need is available within a 15-minute radius. Such attempts to reimagine urban mobility have proliferated in prescriptive narratives of how to build better, healthier, and more sustainable cities to cope with the basic facts that many cities are growing and that traffic congestion and pollution so far have followed in the slipstream. Walking and cycling are promoted in cities around the world through policies as well as designated paths and routes that are implemented as more or less permanent fixtures in the infrastructures of mobility. These reimaginings of urban mobility also imply the necessity for planners to focus on the sensory aspects of cities where 'qualities of space, matter and scale are measured equally by the eye, ear, nose, skin, tongue, skeleton, and muscle' (Pallasmaa, 2005, p. 41; see also Urry, 2012). Movement, in short, has sensory qualities shaping perceptions and practices, albeit these qualities do not offer themselves up to generalisation. At the same time, it is embedded in political and social worlds.

In that regard, movement in the city is about much more than people going from where they are to where they want to go. 'Issues of movement, of too little movement or too much, or of the wrong sort or at the wrong time, are central to many lives' as Mimi Sheller and John Urry note in their summation of what they call the 'new mobilities paradigm' (2006, p. 208). Through such a perspective, it becomes all the clearer how movement is a central characteristic of the city as lived, not least since movement creates social and material interactions, and thus also plays a significant role in shaping the atmospheres of urban spaces.

In this chapter, we follow the broad range of interest in movement in geography, sociology, and anthropology as it points us to how the politics and planning of mobility are entangled in materiality, cultural logics, and everyday life. But our focus here is not so much on mobility, including the details of how it is planned, as it is on the relationship between atmospheres and movement in a broad sense. The chapter thereby explores movement to develop 'an appreciation of the capability of bodies to affect and be affected by the full extent of their environments, and especially by those forces of which we are not always immediately aware' (Simpson, 2017, p. 18). This entails attending to ways of feeling, perceiving, and attuning to one's surroundings, whether when walking (Duff, 2010; Ingold & Vergunst, 2008; Jensen et al., 2021; Middleton, 2011, 2021; Morris, 2004; Pinder, 2011), running (Larsen, 2022), cycling (Aldred, 2010; Simpson, 2017, 2018; Spinney, 2006), driving (Laurier et al., 2008), commuting by train (Bissell, 2009), flying (Lin, 2021), or engaging in various other kinds of mobility (Vannini, 2009).

David Bissell's work, in particular, has foregrounded the affective and atmospheric aspects of mobility and, equally important, the role of stillness and immobility in human lives (cf. Bissell, 2018, Bissell & Fuller, 2011). One such atmospheric aspect of mobility is how notions of comfort are negotiated through mobility (Bissell, 2008, 2009, 2010, 2018). For instance, he shows how a common understanding of commuter discomfort sees it as a *residual effect* of poor infrastructure and thus overlooks how people seek ways of extending their agency through tactics and knowledge about their own bodily capacities and attending to what causes discomfort. Here, '[c]omfort is no longer solely an attribute of an object but more a set of anticipatory affective resonances where the body has the capacity to anticipate and fold through and into the physical sensation of the engineered environment promoted' (Bissell, 2008, p. 1701). In this specific case, comfort is then both a bodily sensation as well as a spatial and social phenomenon that shapes perceptions and practices. What we take from Bissell and others within the new mobilities paradigm who focus on the atmospheric and affective qualities of moving, is the importance of attending to the intersections between the atmospheres, bodies, practices, and materialities wrapped up in mobility, leading us to the question of how movement constitutes the atmospheric city.

We start by exploring how movement is facilitated by design in urban spaces and how this facilitation relates to wider issues of access and inaccess. We then turn to the relationship between attunement and movement, paying particular attention to ways of perceiving and relating to the world through mobility, more specifically through walking. This relationship is then further explored through a look at more accelerated ways of moving through the city, such as cycling, which serve to highlight some of the intricacies of how practices play a key role in shaping the atmospheric city through movement. Aside from movement being entangled in ways of perceiving cities and in people's active practices within them, it is also a central feature of social life as we move among others and find our place in the world. We thus focus on the sociality of movement in the atmospheric city toward the end of the chapter. Or as philosopher Tonino Griffero notes, people approach the city in movement not as experienced through any *one* atmosphere, but as 'always an archipelago of sub-atmospheric isles (city areas and districts, streets and even favourite bars!), a kaleidoscopic space of plural intensities and lifestyles, whose atmosphere deeply changes depending on performative flows of human dynamics'

(Griffero, 2021, p. 43). It is these flows, and the movements that constitute them, that we turn our attention here.

Moved by design

The terrain and historically grounded layout of Copenhagen, Oslo, and Stockholm offer distinct affordances to movement. Copenhagen is flat and relatively dense while Oslo slopes toward the fjord, and Stockholm is characterised by low density and rocky hillsides. One thing the three cities do have in common, however, is that their landscapes are shaped in part by water, with bridges (and ferries) connecting islands and various urban districts across inlets, canals, and streams. In recent decades, large stretches of the urban harbourfront that mark the borders between land and water, have been undergoing development in all three cities.

In Stockholm, neighbourhoods such as Hammarby Sjöstad, Liljeholmen, and, more recently, Stockholm Royal Seaport, have all been high-profile waterfront development projects focused on expanding the residential areas of the city. In Copenhagen, the old post-industrial harbourfront along either side of the body of water that separates mainland Copenhagen from the island of Amager has seen similar large-scale projects take shape in recent years and decades: From Sluseholmen in the south to Nordhavnen in the north. Between these two outer poles, the harbourfront also features high-profile public buildings, from the Copenhagen Opera House, inaugurated in 2005, to the Royal Playhouse, constructed across the water in 2008. Similarly, the developing waterfront at the fjord marking the southern border of the city of Oslo features a mix of newly built residential pockets and iconic public architecture. The imposing Munch museum, whose exhibitions centre on the legacy of Edvard Munch, opened its doors to visitors in late 2021 and is thus the most recent architectural addition to the waterfront district of Bjørvika, which we visited briefly in Chapter 2. As mentioned there, this area also houses the Deichman Bjørvika public library, completed in 2020, and, notably, the Oslo Opera House, which has been a landmark of the area and wider city since 2008.

Specific design interventions and architectural elements added to these landscapes have also come to play a role in how movement itself has been designed for in different ways than in the past. The case of the Oslo Opera House in particular offers some pertinent nuance to issues of how movement and design interplay in the contemporary city seen from an atmospheric perspective (Payne-Frank, 2022). The overall design of the building espouses a kind of 'openness' that comes across simultaneously in public narratives about it, in how its atmosphere is staged through design, as well as how it is used, perceived, and felt among our participants (Payne-Frank & Schwabe, 2022). Perhaps the most eye-catching expression of this openness is the form of the exterior of the building itself, with the roof of the Opera House being open and accessible to the (able-bodied) public year-round. While, as described in Chapter 2, the interior of the house muffles and filters the surrounding city, the sloping roof provides a different perspective on the city and gives it a different feel. Indeed, from the top of the building, the city becomes a panorama to those taking in the view. Meanwhile, many people seem to engage with these immediate surroundings in ways not usually observed in other public spaces, as if the design somehow challenges the usual flows of the city by encouraging different forms of dwelling and movement that would not otherwise be considered appropriate.

At the Oslo Opera House, some people take a moment of stillness on the rooftop to take in the views and impressions of the city, while others slowly stroll up, down, and around the front as might be expected from most such public spaces. Others, however, practice dance routines at the top of the building – with or without accompanying music – and many make use of the descent to roll down on the wheels of scooters and bikes, imitating skiers on a mountain slope. Quite a few people use their visit as a photo opportunity, jumping up and down in front of cameras or lying themselves flat down so as to use the white marble of the roof as a backdrop to their selfies. Runners move through the space, with or without ascending the roof, and tourists are often seen rolling their suitcases along with them as they stop by the roof on the way to or from the central train station nearby. On one occasion, we came across a small group of artists who had staged an afternoon tea party on a blanket covering the marble in front of the main entrance. On another, a party of Buddhist monks were slowly moving through the space taking photos. Sometimes, after dark, the sources of the spotlights that illuminate the very top of the building are located by visitors, who begin to move in playful ways in front of them, casting large shadows as they do so. In many ways, the exterior of the building has become a space for exploring movement, both one's own and that of others. The design – and designed atmosphere – at the Opera House, in that sense, aligns with the more general point that atmosphere 'gives rhythm to our movements and modulates the manner in which we move' (Thibaud, 2011, p. 209).

Figure 4.2 Playing with shadows at the Oslo Opera House.
Photo by Jeremy Payne-Frank.

At the same time, this is not to say anything goes in this space. As has been argued elsewhere in connection with this particular case study, the openness of the Oslo Opera House is partly grounded in 'elements of direction and enclosure while hinting at a certain porosity' (Payne-Frank & Schwabe, 2022, p. 15). Although the design of the space sets it apart and leaves it open to a variety of modes of movement, there is still – and inevitably – something restrictive in the makeup of the space. In a relatively straightforward way, the placement and characteristics of walls, glass panes, and ledges mark which areas are accessible and guide people's movements even as the space *feels open*. In that sense, when looking at movement from an atmospheric perspective, we can begin to identify some possible tensions between how design facilitates, seduces, and directs movement and how moving in a certain space feels, once again pointing to the subtleties of 'ambient power' (Allen, 2006). Indeed, when we focus on the felt qualities of movement in this chapter, aside from the individual corporeal sensations, it is also to highlight how atmospheres 'become active shapers of normativity, expected rhythms and choreographies that necessarily exclude certain bodies, and by extension categories of people that might become inappropriate to these specific branded, circulated, and implemented atmospheric orders' (Kazig et al., 2017, p. 5).

Of course, far beyond Oslo, designers have been playing with ways of encouraging movement in public space for years. Beyond attempts to better facilitate mobility across cities, planners, and landscape architects have been producing public squares and parks that seem to invite users to both move and linger differently from what might have been the norm half a century ago, for instance by making room for a plethora of mobility forms, ranging from walking to skateboarding, scootering, and cycling.

This is also the case at Israels Plads in Copenhagen, which we described in Chapter 2 (see also Bille & Hauge, 2022). The varied elements that make up Israels Plads allow for different rhythms to be present in the space simultaneously and thus hark back to the openness experienced at the Oslo Opera House. There are the usual benches for sitting, but then there are also various paths through the square, the basketball court and its adjacent play area, water elements, and not least the two sets of concrete steps that encourage those who would like to stay seated to do so as an elevated audience to the scenes playing out below them. As the architects who designed the square write:

> The new plaza aims to revitalise this significant urban space and turn it into a vibrant and active urban plaza that celebrates diversity and livability. This is where children play during school breaks and friends meet for a beer after work.
>
> (Cobe, 2018, p. 91)

With all of these design measures, different forms of activities, movements, and lingerings are made possible. In addition, and importantly, different experiences of atmosphere come into play. Indeed, at both the Oslo Opera

House and at Israels Plads, the designed facilitation of movement and stillness shapes ways of being present and attuned in these spaces, also in unexpected ways. This, in turn, shapes how the atmospheric spaces themselves become constituted, not least in relation to notions of openness and closure.

Looking at the design of movement and its related restrictions leads us to consider not only how cities feel but also to pay attention to how cities can work as systems 'of access and *in*access' while being inherently atmospheric. This raises the question of how 'does it feel to move, to be moved, or to be blocked from moving' (Jensen, 2022, p. 39; original italics; 43)? In these cases, it was a seemingly free and open atmosphere facilitating movement that shaped the identity of the place.

Perceiving through movement

In her book on finding a sense of belonging upon returning to her native Kentucky, bell hooks writes:

> Searching for a place to belong I make a list of what I will need to create firm ground. At the top of the list I write: 'I need to live where I can walk. I need to be able to walk to work, to the store, to a place where I can sit and drink tea and fellowship. Walking, I will establish my presence, as one who is claiming the earth, creating a sense of belonging, a culture of place.
>
> (2009, p. 2)

As a mode of finding a place to take up in the world, of establishing a sense of belonging, walking plays a central role in how people relate to their surroundings, whether in cities or in more rural settings (Jensen et al., 2021; Middleton, 2011, 2021; Pinder, 2011). More than simply a means of transportation, bell hooks reminds us that we become established in place at least as much via moving *through* it, as we do by staying *in* it and building a more static, material home around us. Alberto Pérez-Gómez similarly makes an important point in his defence of walking, namely that it is 'emblematic as a primary mode of perception', not least 'because it provides a space of recollection and meditation' (2016, p. 4), echoing Kim's statement earlier in this chapter.

In this section, we turn to the role of movement as a mode of perceiving and encountering the city. In doing so, we home in on walking as a form of mobility that often spurs people to attune to their surroundings in ways that accentuate and fine-tune their perception. However, as pointed out above, we do not seek to offer a generalised view of the qualities of walking for putting people in a meditative state (or any other form of movement for that matter). Rather, walking is presented here as simply one way of fostering perceptions of the atmospheric city through movement.

While being interviewed over the phone, Thomas, whom we met in Chapter 2, took a walk in his neighbourhood of Vesterbro in Copenhagen. Upon walking down Istedgade, a central and often quite busy street, he said he was surprised at how many people were there, spending time outside. This was early on during the pandemic in 2020, a time where, at least in Denmark, contagion was seen as tactile, transmitted mainly via touch, and not air as was later established. It was spring in Copenhagen, and the first lockdown had only been in effect for about a month. Still, Thomas' expectations had already been adjusted to fit a situation where public life – both in actuality and to his imagination – had diminished drastically. But this day was different. 'Suddenly Istedgade feels like it used to, full of people everywhere', he said. Already having adjusted to constantly thinking about keeping his distance from other people, Thomas felt it a 'bit strange' that the streets should be so busy in a context where contagion was very much a concern. At the same time, he explained, he could understand why all these people were out and about. As he put it, 'I get it, because what drives people out is the same as what drives me out: We need air, we need to be outside because otherwise we would go crazy in our apartments.' In this instance, moving out of one's home and into the space of the city was not primarily a means of transport, it was a means of coping mentally.

At the same time, walking along a busy street and being around others as they moved offered Thomas a mode of perception by which he took in the atmosphere of the pandemic city. Although things seemed almost ordinary in Thomas' experience on that day, he also remarked that he noticed far fewer cars and bicycles in the streets, and less traffic in general than before the lockdown. To him, as he went for his neighbourhood walk, this allowed him to perceive his surroundings in a new and different way. Being outside was no longer about working his way through traffic to get somewhere. To Thomas, going outside had become about noticing things he had not really paid attention to before: The wide sidewalks of certain parts of Copenhagen and the smoothness of the broad bike lanes that ran parallel to them. The COVID-19 lockdown had taught him that when you just walk around without having somewhere to *be*, 'when you're just glad you get to go outside', you simply notice things more; not necessarily things that appear new to you, but things that appear to you in a new way.

What Thomas and Kim related to us shows the way movement – and sometimes certain types of movement, in this case walking – can impose on people a sense of heightened attention, making the very act of moving a mode of sensing and perceiving the city anew. As Ole B. Jensen, Michael Martin, and Martin Löchtefeld (2021, p. 3) note, 'walking is perceiving (...) To walk is, in this sense, to register the world and to perceive the environment in an embodied and specific way.' To that we might add that this point could be expanded to include how other forms of mobility allow for similar or yet entirely different ways of perceiving the world.

Figure 4.3 A solitary walker in Stockholm.
Photo by Siri Schwabe.

The experience of becoming attentive to one's surroundings through movement was a recurring theme among people we spoke to, who often remarked that they simply notice more of their surroundings, especially when they walk. Like Thomas and Kim, people experienced something akin to what has been described as being 'affected by micro-climate, smells, noise, sights, the presence of other bodies as well as vehicles in the urban spaces' (Jensen et al., 2021, p. 4; see also Ingold, 2015). Attending such experiences of movement in the city illustrates the porosity between 'a moving, sensing subject and an inanimate, material object-world' (Jensen et al., 2021, p. 4). Noticeable is, however, that there is a danger of romanticising or making walking more 'authentic' or 'real' than other embodied mobility practices (Jensen et al., 2021, p. 3). Walking may also be a strenuous form of exercise and can indeed be impossible for people with certain disabilities. Meanwhile, the examples presented here illustrate how one of the capacities of movement is to allow people to see the city anew, and that this capacity is often most clearly experienced when moving on foot.

When we move through space (whether on foot, on a bike, or in a vehicle), our bodies engage with our surroundings in ways that allow us to pick up something of their atmospheric qualities. As Tim Flohr Sørensen notes: 'Movements do not just result in a transportation of bodies, but have the capacity to produce sensibilities, generative of the very experience of a room or a situation' (2015, p. 66). In more basic terms, the way we move shapes our ways of perceiving our surroundings. This means, he argues that 'sensing one's body in its field of movement is also a feeling of one's extension and magnitude of movement, a feeling of spatial constraint, intimacy, or exclusion' (Sørensen, 2015, p. 66). Moving, in that sense, may get people in a flow, heighten their attention, or make their bodies contract, hurt, and become fatigued, while it fundamentally shapes their perception of the environment.

Movement through space, no matter through which medium and at what pace, always carries a double potential; that of (eventual) arrival and that of journeying. Reflecting on the multiple ways and implications of walking, Rebecca Solnit (2001) distinguishes between wayfaring and wandering, positing that the wayfarer simply makes their way from one point to another, while the wanderer moves more slowly and without a set destination. Through movement along the ground and through atmospheres, walkers – and sometimes those who move in other ways – gain a knowledge of their world that is 'open-ended and exploratory' (Ingold, 2015, p. 48, see also Ingold & Vergunst, 2008). Unlike the wayfarer, always aiming for some destination or other, the wanderer here employs a different type of motion; one that allows for an almost meditative relation to the city through which someone becomes lost in thought, watches the sunset, stops and rests, or just meanders aimlessly, taking in the urban environment and establishing their personal presence within it while doing so. Perhaps especially, although not exclusively, when moving on foot, people become attuned to both themselves and the city in ways that highlight a heightened sense of taking in – indeed encountering – the city anew. As Cameron Duff summarises: 'To walk is to be affected by place and to simultaneously contribute to the ongoing co-constitution of self and place' (2010, p. 887).

The focus here on encountering and perceiving the atmospheric city through movement, and particularly walking, may appear to evince a strictly phenomenological approach devoid of attention to issues of power and politics. However, it is evident that the ability to walk or wander the city is also based on planning decisions and design, not unlike what we saw at the Oslo Opera House and Israels Plads above. In that sense, achieving a *15-minute city* or a *walking city* is a political choice. The possibility for walking, as much as other forms of mobility, does not come effortlessly in cities, but is the mark of a material infrastructure that offers certain ways of moving and thereby ways of perceiving the city to and through certain bodies.

Moving in practice

Earlier in the book, we looked at how people actively seek out atmospheres and put in work to attune themselves into existing atmospheres or orchestrate ambient settings to fit a certain need or desire. In this section, we extend that focus on the practical aspects of atmosphere by delving into how practices related to movement shape the atmospheric city and people's experiences of it. While the previous section dealt with how cities appear to people in perception as movement occurs, this section presents a subtle shift in perspective through which movement is approached as a form of activity. It remains that our considerations do not limit themselves to one mode of movement in particular. However, we rely mainly on examples centred on cycling in this section; not because other forms of moving do not entail active practice that have bearing on how cities feel, but because the experiences of cycling that people have related to us are especially illuminating of how atmospheric cities come about as cities in and of movement.

Of course, when observing the traffic in the busier areas of Nordic cities, we see an increasing number of ways in which people get around. There are the slow strollers, the fast-paced walkers, the joggers, the runners, the cyclists riding a range of different types of bicycles at different speeds, the bus, tram, and train passengers, and the drivers of cars, trucks, and public transportation vehicles. Moreover, electric scooters have at times become an almost omnipresent mode of transportation for tourists and locals who speed down sidewalks, bicycle lanes, and streets. Around schools and similar institutions, children and young people kick their way along on non-electric scooters. More seldom, but not unseen, are people on skateboards, rollerblades, or riding various forms of human-powered bicycle taxis, hoverboards, and similar innovations in the field of mobility. In Copenhagen in particular, cargo bikes – like the one on the front cover of this book – take up a lot of the already busy bicycle lanes while further north in Scandinavia, roller skis are not an uncommon sight.

Perhaps especially when moving at a faster pace than possible by walking, our surroundings appear to us somewhat hazier. When running and cycling on a busy pedestrian walk or bike lane, people's attention must necessarily to a greater extent than on foot be focused on what is beneath their feet and wheels, what is coming up ahead, and whether there are potential obstacles to their movement around them. As a counterpoint, if running in little traffic or cycling along a nature trail, they may become more absorbed and introspective. As Pallasmaa notes, perception of the world is fragmented, it is 'held together by constant active scanning by the senses, movement, and a creative fusion and interpretation of our inherently dissociated percepts through memory' (Pallasmaa, 2014, p. 243). That, of course, also changes how people become attuned to a hectic rush hour, a late evening with no people on the street, how they sense the atmospheres around them, and, significantly, how they contribute to constituting them in practice.

Figure 4.4 Cycling in Copenhagen.
Photo by Mikkel Bille.

In Copenhagen, 'bike culture' has long been a staple of narratives of the city, both officially and unofficially. As Jonas Larsen (2017) notes, 45% of Copenhageners cycle to work or their place of education in the city. Metrics being what they are, however, this does not qualify what 'cycling' feels like, as Larsen also illustrates. Cycling in the rain or sleet is vastly different from riding along in the sunshine, just as it is the case that experiences and practices change and are adjusted to the weather while walking or driving. A Monday morning during rush hour on one of the main roads with thousands of other commuters on bikes is vastly different from a slow afternoon sightseeing tour along the backstreets of the city. The feeling of cycling through the old inner city with its stretches of cobblestones is notably different from how it feels to speed along on a paved bike lane passing rows of concrete housing blocks. Cycling, in other words, is another in a range of practices of movement that embeds people in shifting atmospheres; a practice that attunes the body and through this also changes and charges atmospheres.

Local topography and culture play an important role in the popularity of different movement practices. For instance, cycling can be a mode of transport, an accessible form of exercise, or a serious sport depending on the context, including the landscape (natural and built) and established ways of doing things. Cycling in Stockholm, with its hills and abundance of water, is different from cycling in Oslo. Meanwhile, in a topographically flat Copenhagen, it is often simultaneously a cheaper, faster, and more flexible solution than public transportation or cars. Here, it is a practice both imbued with meanings and less physically strenuous than it would be in a more hilly city. One participant in Copenhagen noted how, to her, 'bikes are freedom'. They allow her to go fairly quickly to places in the city and can likewise be brought on the train when she feels like going for a ride even further away. Cycling, like driving and walking, from an atmospheric perspective, is both a form of movement as well as a felt and social practice. To the person above, the experience that 'bikes are freedom' is very much tied to how she experiences the very practice of cycling and how cycling offers her a vantage point from which she can take in the city and its surroundings actively, through practice.

Many participants in Copenhagen talked about cycling and the particular atmosphere it creates in the city simply due to how common a practice it is. One person, in particular, spoke vividly about cycling. Robert was in his mid-20s and worked in a kitchen at a school about half an hour's train commute outside Copenhagen. He was living in the Northwestern part of Copenhagen, known for being a fairly rough part of town, but he kind of liked its 'chaotic atmosphere' – its clutter, diversity, and being a little less polished than other parts of the city. Since he lived a bit away from the city centre, he relied in great part on his bike to get to the main train station from where he would take the train to work. As he explained to us, he cycled both because it is a flexible mode of movement that left him without having to worry about bus schedules, but also, he said, because it was possible to escape from unpleasant situations by bike.

Indeed, he had experienced situations when riding on trains, where he could not simply leave. When he takes the train home from work in the late evening, he told us, he often feels uncomfortable with what he perceives as a very 'macho atmosphere', with young men agitating each other and looking as if a fight could be imminent. He often moves to another compartment when this happens, but if that is not possible, he stays and remains uncomfortable for the length of the trip. In that respect, he highlights one of the downsides of public transport – particularly to women, minorities, and vulnerable people – namely that it is by many associated with feeling unsafe at train stations, uncomfortable waiting at a bus stop, or the corporeal feeling of being pressed up against other bodies during rush hour (Ceccato, 2020; Nourani et al., 2020; Shibati, 2020). In contrast, when cycling, he felt there was a lot of space in Copenhagen, and, with that, less of a risk of exposing himself to discomfort:

> So I think if there's just a little bit of sunshine, then it's always very liberating to cycle through the city and cycle really fast because you can feel that you're getting the wind in your hair. And also that you can follow what happens around you. So, you just hear someone standing and selling strawberries on the street corner, or someone discussing with their

> child that the child should not have candy. You capture these fragments of conversations and what's happening around you. So, there are a lot of sensory impressions that you miss if you go down to the Metro, for example, where it's very sterile and dark and closed off. Nothing else happens other than people going from A to B in the Metro. Whereas cycling, and to some extent also taking the bus, is an experience of things around you, and you see the living city [*den levende by*]. Does it make sense? […] and of course you experience more on a bike because you are out in open space [literally 'free space' in Danish, *frie rum*]. And you can better hear what's happening around you and have all the senses activated – you can also smell and things like that. You can feel the city in a different way.

What Robert described here was clearly about how his perceptions of his surroundings were shaped through movement. The point here, however, is not so much that movement prompts a heightened sense of the urban environment – although, as we also saw above, it does – but that the experiences Robert describes are made possible through practice and choice of means of movement. On trains and in the Metro, Robert finds himself in spaces that are not open and 'free' and where his own capacity for active movement is restricted. In contrast, when cycling, he is able to move at his own speed and leisure, to remove himself from unpleasant situations if they should arise, and to take in the city in a way that allowed him to feel it as a 'living city'.

We also spoke to Jens, whom we met in Chapter 3, about how he explored the city from his bicycle. From the seat of his bike, he explained, 'you experience many different parts of Copenhagen, many different neighbourhoods that have certain characteristics'. Continuing, he added how 'in some places you sense that there are big apartments and relatively big cars. And in other places [you see] smaller apartments and lots of bikes', showcasing 'different parts [of the city] and different ways of life, which are reflected in the neighbourhoods.' To Jens, who greatly enjoyed his two-wheeled tours, cycling had become a way for him to bring the city together into a potpourri of impressions, bringing the micro-characters of neighbourhoods together and into a sense of the macro-character of the city. Significantly, the neighbourhoods he moved through did not just *look* different to one another, they also *felt* different. Angharad Closs Stephens has made a similar point:

> The atmospheres of urban life are mostly multiple: we might traverse back and forth between several insides and outsides in a very narrow timespan. And we might only realise a change in atmosphere *after* having experienced it – that is to say, it is not necessarily *visible* but might be better captured as a *feeling*.
>
> (2015, p. 100)

Moving through the city on his bike, Jens experienced these changing neighbourhoods in a way that was clearly shaped by the kind of 'unconscious and unfocused peripheral perception' (Pallasmaa, 2014, p. 38) that often seems to be what characterises people's perceptions of atmospheric spaces. On his trips, he would feel the rhythms of the city, the pace of his bike, and the temporalities of everyday life (cf. Jensen et al., 2015; Lefebvre, 2004). This is not to say that cycling does not offer time for contemplation, while walking does. It is rather that the attunement to the atmospheric city unfolds through people's bodily engagement with it, through practice. To Jens as much as Robert, it was his own active movement that facilitated a process of attunement to the different atmospheres of his surroundings and, as he moved, he also became part of the flow of the city, part of what makes Copenhagen a city marked by 'bike culture'.

Moving among others

While the section above focused on the practices of movement, these practices are also entangled in social life. In Solnit's (2001) extensive ruminations on wandering, she draws on the example of Søren Kierkegaard and paints a picture of him as one among several preeminent philosopher-walkers. According to Solnit, Kierkegaard likened his walks to 'botanizing', but, as she points out, 'human beings were the specimens he gathered' (2001, p. 23). She describes how, for Kierkegaard, strolling along the streets of Copenhagen 'was a way to be among people for a man who could not be with them, a way to bask in the faint human warmth of brief encounters, acquaintances' greetings, and overheard conversations' (Solnit, 2001, p. 24).

Besides using walks as a means of thinking and working through ideas, for Kierkegaard, a man who did not have much – if any – of a social life at home, the streets became his prime space of human encounter. When strolling down streets or through urban parks and similar spaces, people are rarely alone. For some, like Kierkegaard, part of the attraction of going out for a walk is indeed the possibilities it offers for interacting – or perhaps more often simply observing and sharing a space – with known and unknown neighbours and visitors. During strolls close to home especially, people are also often faced with local personalities: The older woman whose name remains unknown but whose habit of wearing only yellow makes her stand out on the street, or the habitual drinkers on the corner, or the man who is known to feed the birds in the park every day. But even on the outskirts of the Scandinavian cities that we focus on in this book, walkers and other urban movers rarely go for long before being faced with people or reminders of the human lives playing out in their vicinity. A dog walker and their companion might cross their path or lit-up windows along the facades of apartment blocks might reveal something of the people that live inside. As Tim Ingold and Jo Lee Vergunst note, 'Not only, then, do we walk because we are social beings, we are also social beings because we walk' (2008, p. 2).

As 'always already attuned', wayfarers and wanderers, walkers and cyclists, passengers and drivers, all move through the atmospheric city and are parts of it, following and establishing flows and norms of how to move – or ways of diverting them. In other words, different forms of mobility facilitate different engagements with atmospheres, but all involve a process of attunement that may be shared across mobility forms. When participants feel focused, if slightly overwhelmed, in the bike lanes of the early morning rush hour, they are no more or less attuned than they are on the pavement, strolling slowly along on an afternoon walk. Of course, these two ways of attuned movement are not the same. However, they are both ways into, as much as through, the city.

Figure 4.5 Sergels Torg, Stockholm.
Photo by Siri Schwabe.

In large cities, sometimes people find their surroundings moving at a pace quicker than their own; when people move slowly and quietly, and perhaps desire stillness, but those around them seem rushed and navigate the world by an intensified flow. Some of our cycling participants would, for instance, avoid main roads with lots of fast-paced traffic because of their hectic atmosphere, even if there were bicycle lanes for safety.

The tensions between fast and slow movement also became a topic of conversation when we spoke to Axel on a cold winter's afternoon at the central square of Sergels Torg in Stockholm. Besides being a busy public space in the heart of the city, adjacent to the main shopping street and just a stone's throw from parliament and the royal palace, Sergels Torg is also a traffic hub: Through glass doors at one end of the space, people enter and exit the bustling metro system and find access to the nearby central train station. To Axel, it has 'a charm even though it's not charming. I like the brutality of it, but otherwise I think it's one of the most disgusting (*vidrigaste*) places in Stockholm.'

Sergels Torg was inaugurated in the 1960s, at a time when much of central Stockholm was undergoing changes. This is evident in the square's surroundings as well. In contrast to nearby Gamla Stan – the old town – where everything is kept almost exactly as it was a few hundred years ago, the buildings around Sergels Torg bear the architectural marks of having only come up a few decades ago. Like other areas developed during the same period, Sergels Torg features a type of urban planning whereby traffic is ordered in horizontal layers. In this case, motorised traffic is relegated to the upper level while pedestrians walk down below, out in the open across the square itself or underneath the road along parallel tunnels that act as shopping arcades and have several different entry and exit points.

The fact that a large proportion of those who work, study, and/or live in Stockholm pass through this traffic hub daily means that it is almost always busy. On this day, a snowy and cold Monday in the late afternoon, most people passing through seemed in a rush, many of them walking quickly with headphones on and carrying backpacks or briefcases. Intermittently, someone took a cigarette break, often as they stood against the wall of the metro entrance and relatively often, individuals or small groups lingered for a couple of minutes as they waited for whomever they had plans to meet up with. Only three groups of people seemed to stay in the square for more than a few minutes: A small group of Roma men and women; the police officers inside the vans that were parked for shorter or longer periods in the centre of the square or on the street above as they kept the square under surveillance; and the three Jehovah's Witnesses who had set up a temporary stand from where they held out pamphlets to passersby. To Axel, all these impressions felt stressful.

> There is no calm here. It's not that people here are unpleasant, but I wouldn't want to hang out here, I would get stressed. This is a place that you just want to pass through. There are a lot of people, and a lot of people just going back and forth, and I get restless from that.

Axel's experience at Sergels Torg underscores two important factors forming the relationship between movement and the atmospheric city. Firstly, while mobility practices can lead to experiences of freedom as people make active decisions on how to move – as in Robert's case – these practices are often ordered and restricted by available infrastructures in tandem with the basic condition of (urban) life that people often need to be in a certain place at a certain time. To Axel, Sergels Torg was a place that he needed to pass through, but that simultaneously stressed him out. Secondly, how cities feel relies not only on people's own movement within them, but very much hinges on the movements of others; in this case, the hectic and fast-paced movements of people passing through the square in a rush, leaving Axel feeling 'restless'. Here, Axel was subjected to a flow that was far beyond his control and this, in turn, affected his overall sense of the place. In sum, movement shapes the atmospheric city not least through planning and design as well as through the intensity of its very flow.

Importantly, not only social encounters and experiences of flow but also someone's position within the social worlds of the city plays a role in how they move through it, and how their movements play a role in attunement. Those who make their way around public spaces after dark, for example, often find themselves confronted with their position as atmospheres change with the light conditions (cf. Edensor, 2017; Hvass et al., 2022; Oznobikhina, 2021; Yang et al., 2022). On an early winter's evening, while walking through the central train station in Stockholm, we spoke to a local woman in her 50s. She told us that she had just left her abusive partner a few months prior and had been without steady accommodation since then. She was constantly wandering in the city. When we asked her where she had spent the previous nights in the extreme cold, she told us that she had simply been 'nightwandering' (*nattvandrade*), trying to keep warm.

She went on to explain how 'scary' it had been and how it tends to get particularly bad when the lights go out in the buildings and the city gets darker and feels emptier. She had tried to rest in a stairwell the night before, close to the station where we met her, but had been discovered and kicked out. Now she was trying to scrape together some money to get into a shared dorm in a nearby hostel where she had stayed before. Evidently, to this woman and others, walking is not merely a leisurely activity, nor is it about transportation. It can also be a mode of movement so infused with social vulnerability that its other capacities become much less pronounced.

As several scholars of the night have shown, nighttime mobility holds a variety of connotations and potentials, which are woven into political, economical, and social imaginaries (Beaumont, 2015; Dunn & Edensor, 2022; Shaw, 2018; Van Liempt et al., 2015). At the same time, with the different flows of the dark hours, from winter rush hour to dinner time and into evening and night activities, it becomes clear that to talk about the night in the singular omits that in lived cities, there is a plurality of *nights* marked by a multiplicity of experiences, practices, and temporal horizons.

Indeed, the specificities of someone's social position within the city and their related possibilities of movement speak volumes about the stark differences in how both the atmospheric city and mobilities within it are experienced. Often, as in this case, walking as a mode of movement is not a privileged choice but a result of particular socioeconomic conditions. Indeed, walking – especially at night – has historically been seen as an activity for the marginalised. As Matthew Beaumont describes in his book on nightwalking in London, only those perceived as somewhat seedy characters would find themselves out at night, loitering and strolling through the streets. As he writes: 'Who walks alone in the streets at night? The sad, the mad, the bad. The lost, the lonely. The hypomanic, the catatonic. The sleepless, the homeless. All the city's internal exiles' (2015, p. 13).

As the examples above foreground, movement in and through cities is an inherently social phenomenon that is marked both by the presence of others and by one's own presence relative to others (Cf. Middleton, 2021). In that sense, how cities in motion feel is as much about how it feels to be around others as it is about how people's own movement shapes perceptions and works to attune them in practice.

Conclusion

As we have called attention to in this chapter, the atmospheric city is a city in movement. Ash Amin and Nigel Thrift note how

> cities exist as means of movement, as means to engineer *encounters* through collection, transport and collation. They produce, thereby, a complex pattern of traces, a threadwork of intensities which is antecedent to the sustained work of revealing the city minute on minute, hour on hour, day on day, and so on.
>
> (2002, p. 81, original italics)

What we are dealing with is more than merely a movement of people and physical objects, but also an affective movement that emerges through processes of attunements to atmospheres. Overall, movement in and through atmospheres helps constitute what the city *is* in an atmospheric perspective; what it is as a designed environment, what it is as an environment that is subject to perception, what it is as an environment felt and shaped through human practices, and what it is as a social environment.

People get a feel for their surroundings – seek to attune to them – through movement. They perceive the world through movement and use movement actively to enhance both perceptions and overall experiences. But, no matter how people choose to move, they are also continuously exposed to surprising elements and flows that the city throws at them on any given day as they move through it. Moreover, the particularities of how people move through space as individuals with discrete bodily abilities and traits affect perceptions and practices simultaneously. With that, attending to the atmospheric qualities of moving through space in its nitty-gritty details reveals a great deal about how people perceive and actively move within urban environments, which are also inherently social spheres, beyond what metrics can offer. Indeed, it allows us insights into the relationships between motion and processes of attunement in atmospheric cities that are also spaces of porosity, of ongoing motion.

Taking movement seriously from this perspective involves thinking beyond the human scale and shifting the focus from human-centric design focusing on the human body or scale to a life-oriented approach to developing cities, where there is a recognition of the need for space for introspection, for being alone, comfort, and for moving fast or slow, beyond the common categories of safety and security. Put differently, movement does not only refer to bodies in motion through space but simultaneously to the relations that are forged through movement. Notably, these are relations between people and other people, relations between people and non-human presences, including the material world they move along – paved roads, buildings, vehicles, city nature – as well as relations between people and places, broadly understood, that harbour atmospheric qualities and that have a certain feel to them. This movement, we suggest, also points to the porosity that marks atmospheric cities in that all the presences that shape it continuously seep into each other, not least via tangible forms of movement such as the ones we have explored here.

Figure 5.1

5 Cities of care
Nurturing atmospheres

> I feel safe in Nørrebro. I mean... There are so many different people, and people are so different and lead so many different lives that you end up blending in in a way, whereas in other parts of the city you might feel like an outsider or that you are somehow more noticed.

Kim, whom we met in Chapters 2 and 4, is explaining to us how different parts of Copenhagen engender distinct experiences of homeliness, safety, comfort, or their opposite: An intensified awareness of sticking out and not being looked after. As a non-binary person, Kim describes how walking down the street in Østerbro – a more affluent and less diverse part of the city – leaves them feeling as if 'people are looking a bit like, "wow, you look strange". Or if I am holding hands with my partner while we're walking, people stare.' In contrast, Kim explains, people in Nørrebro react to them with a 'oh, ok... well, I'm just busy with something else', as if their presence warrants no particular attention.

To Kim, there is comfort in being seen but not standing out. Indeed, there is recognition in how they experience being seen in Nørrebro; a recognition that comes with interpersonal responsibility and even protection. As they continue:

> People look out for each other in Nørrebro, I feel. Whenever I've experienced someone shouting at me in the street, if it's in Nørrebro, you can see how people watch and assess the situation; 'do you need help? Or are you ok?' Whereas if you experience someone being unpleasant to you in Indre By (downtown), no one does anything, even if there are a thousand people around and it's right by Nørreport [train station] and everyone sees it.

In looking for a possible explanation to this difference in behaviour and experience, Kim suggests that 'perhaps there's not the same feeling of Indre By being a local environment' while in Nørrebro 'people who live there feel that "this, this is my place; no one gets to bother anyone where I live." So in a way people take care of each other there because it's a kind of village.'

DOI: 10.4324/9781003379188-5

In this chapter, we look at the role of care as an aspect of the atmospheric city. While care has received increasing attention within urban studies in recent years (see Davis, 2022; Fitz & Krasny, 2019; Tronto, 2019), we focus specifically on what Kim expresses so clearly, namely that there are significant experiential links to be found between how a place feels and how it is both perceived as caring – not least through the recognition of local people – and cared about. In the following sections, we first look into what a city of care might mean in the first place. We then look at how urban design plays a role in constituting spaces that are perceived as either caring or uncaring; in other words, we ask how care appears variously as present or absent in urban planning, materiality, and management, and how this influences people's experiences of cities. From there, we explore what it means to live in the atmospheric city for people who play active roles in giving, receiving, and negotiating care, both in relation to each other and in relation to their wider surroundings. This involves considering two pillars of care: Comfort and compassion. It likewise leads us to considerations on the significance of knowledge and familiarity with places in experiences of care. Finally, we explore plurality and difference in relation to care, asking questions about how different people with different bodies inhabit different spaces differently. Importantly, we use the term comfort to denote both an embodied sense of *being comfortable* and a sense of *feeling reassured* or *soothed* (see also Bissell, 2008, 2018).

In the Nordic context in particular, with the emphasis on the welfare state and Nordic design tradition, looking at the city as atmospheric entails paying attention to not only how urban design impacts how care is facilitated and felt, but also how the everyday workings of the welfare state interplay with perceptions and practices of caring and being cared for, not least in the context of 'livability' (see, for example, McCann, 2007; Simpson, 2019). Apart from the importance of funding decisions, urban development projects, and the everyday maintenance of public space, wider institutional structures cannot be disregarded as a crucial aspect of how caring cities are built. Access to housing, healthcare, and education are deciding factors when it comes to who lives in the city and how. They also play a role in how those who live in or otherwise use the city take part in shaping it. Importantly, the presence of a large and strong public sector in the Nordic countries, leaves a mark as municipalities play a significant and very active role in taking care of the streets and squares of the city. Among other things, that usually means that sidewalks are kept relatively garbage-free and that the grass in the public parks is regularly cut. It also raises interesting questions regarding responsibility: If municipal workers are being paid to sweep streets and empty public bins, wherein lies the responsibility to care among ordinary urban dwellers?

In many ways, this chapter builds on the discussions presented in previous parts of this book. Care in the atmospheric city is indeed a phenomenon wrapped up in the social workings of atmosphere, experiences of the urban environment, as well as where and how people move in cities. Care, in that sense, is not simply another mode of engaging the atmospheric city but is perhaps better understood as an overarching quality of that engagement; one that might also benefit from further attention from urban authorities.

Care in the atmospheric city

In July 1995 a heat wave struck the US city of Chicago, leading to a death toll of more than 700 people – mainly elderly and poor residents of the city – over just five days. Scrutinising this event, Eric Klingenberg (2002) argues that a 'geography of vulnerability' linked to race, place, and particular ways of life was at play. In particular, he investigates how the death tolls of two separate neighbourhoods were markedly different. In North Lawndale, home to many African Americans, death tolls were high compared to the nearby, and demographically similar, Little Village, which was largely populated by Latino residents. This difference, he argues, could not be explained with reference to cultural ideals of respect and care for the elderly alone. Rather, he shows, Little Village was a more vibrant neighbourhood, more densely populated, and more commercially developed than North Lawndale, where the de-industrialised and sparsely populated area discouraged elderly citizens from leaving their homes in search of air-conditioned spaces. Here, people did not trust others on the street or even their neighbours to care for them. Lack of air-conditioning and the fear of crime meant that people had a greater propensity to stay inside their overheated homes. Adding to this was also a structural element of a spatial mismatch between the most vulnerable people and the location of medical facilities making response time extensive. Vulnerability and care, in that sense, are at once a social, spatial, and infrastructural concern. As this case so starkly shows, how places outside the confines of the home *feel* may quite literally be a matter of life and death.

When we think of care, we tend to think of tending to loved ones or others who rely on our support, making sure their needs are met and their well-being ensured. We also often encounter the phrase 'to care about' in day-to-day life. It is a commonsensical way of expressing emotional engagement as well as the willingness to direct our resources – of whichever kind – towards people, things, political causes, organisations, and even places. But beyond the usual conceptions of care, what might the term entail, especially when applied to our relationship with cities?

In a very general sense, we might start from a conceptualisation of care as 'a species activity that includes everything we do to maintain, continue, and repair our world so that we may live in it as well as possible' (Fisher & Tronto, 1990, p. 40; see also Conradson, 2003; Lawson, 2007). Following Joan C. Tronto (2015), care is simply everything that goes into keeping the wheels of the world turning; from seemingly trivial activities such as doing the dishes, shopping for food, and making the bed to showing the people around you that they are taken care of by setting the table for families or guests, cooking them a hot meal, and offering them a comfortable place to rest. When people, say, prepare dinner for those they share a table with, they not only look after their nutritional needs but also signal to them that they are there *with* them, *for* them. They, on some level, implore them to feel good.

In that sense, care is about attending to affective connections as much as it is about attending to tangible needs. This, we argue, is also the case at the level of a neighbourhood or a city as much as at the level of the family home; care is about much more than fixing potholes and setting up adequate street lighting so as to meet the basic needs of cities and their inhabitants. It is about how cities might implore their inhabitants to feel good and, in turn, how these inhabitants themselves might play a role in *making the city feel good* (see also Amin, 2006). From such a perspective, it could even be argued that care is also a material relation in that attending to and caring for the city's spaces and infrastructure is also caring for people. For instance, Sumartojo describes how a design workshop in Melbourne showed that street lighting design 'did not need to be large-scale, expensive or dramatic to appear caring. It did, however, need to appear empathetic, sensitive and purposeful to help support feelings of safety' (2022, p. 122).

In a general sense, care can be seen as a process involving four phases (following Fisher & Tronto, 1990; Tronto, 2015). First, care involves *caring about*, which implies the ongoing consideration and identification of needs, reflecting on the state of one's human and nonhuman surroundings. Secondly, care is also about *caring for*: Accepting responsibility and potential role in meeting certain needs, recognising that something has to be done to ameliorate a situation. The third aspect of care in this model is *caregiving*: The actual work that goes into care once a need has been identified and a response has been deemed necessary. Those who carry out this labour are not necessarily the same people who identified the need being addressed to begin with, or indeed those who identified a fitting response to it. Indeed, in cities and elsewhere we often find a stark division between those who initiate practices of care (e.g., politicians, planners, and designers) and those who actually carry out the labour of care (e.g., workers and unpaid members of the public). Fourth, and finally, care involves *care-receiving* as those who gain from caregiving have their needs met and respond accordingly, often in ways that lay bare new or recurring needs, such as when the addition of a public dustbin by a park bench calls attention to the need for more dustbins, or perhaps for a better type of dustbin that will keep seagulls from rummaging through the waste in search for something edible, leaving the space feeling dirty or neglected. From this perspective, care never comes to an end; addressing one particular need or improving one aspect of the environment are simply steps along a path that stretches out ahead with no definitive end in sight.

Importantly, and following this line of thinking, care is both rational and affective; it is both something thought, reflected, and decided upon, and something that is felt (Tronto, 1993). On the surface level, care might seem almost synonymous with the idea of concern; caring about someone or something usually involves being concerned about them or it. It is what people experience when they find that the well-being of others concerns them, touches them, reaching into their lives somehow. Similarly, to say that something does not concern you is to say that it has no bearing on your world and that you, therefore, do not recognise or intend to respond to any needs related to it. However, in certain ways care is distinct from the notion of concern in that it often tends to denote active engagement (Bellacasa, 2017, p. 42). In other words, while concern can manifest itself in a rather passive way, care usually entails doing something to or with someone or something, or at least, following the model above, realising that something *should* be done. In this understanding, care entails practice; indeed, it entails an ethical practice (cf. Tronto, 1993).

However, if we are to understand care as simply an activity, a practice that can be carried out, perhaps on the basis of an emotional bond or other obligation, what about the kind of experience Kim described above? What about the *feeling that* you are cared for in a particular space by whoever – and even whatever – is around at the time, even when no one is really *doing* anything? More specifically, how can we make sense of care as something embedded within urban atmospheres and perhaps even vice versa? From an atmospheric perspective, we might argue that, even when care is not expressed in concrete actions, people are never entirely passive in the world, nor is the world passive as it appears to them (cf. Bellacasa, 2017; Sumartojo, 2022). In simple terms, care can then also be understood as something that is felt, even if only as a potential that is present in certain atmospheres that people pick up and contribute to producing.

With that, looking at care through the lens of atmosphere requires a certain level of flexibility in how we understand the term. If care is something experienced in subtle ways that cannot be explained only through specific practical engagements with people's surroundings – both human and non-human – then we need to think beyond care as simply a mode of activity. A city marked by care is thereby not just predicated on the presence of shiny sidewalks, available seating, and parks that are free from cigarette butts, empty bottles, and other types of discarded waste. It is also built from particular atmospheres that facilitate a sense of comfort, compassion, and mutual commitment, and is consequently anything but static. As Sumartojo and Pink (2019, pp. 81–86) show, a simple piece of mobile equipment such as a trolley for serving meals in a hospital setting is more than simply a material object. Rather, it has physical affordances that signal and facilitate care. Moreover, in the particular ward of their study, where people had limited mobility, it was 'actually central to how an atmosphere of care could emerge as staff had to go to patients rather than the other way around' (Sumartojo & Pink, 2019, p. 85, see also Sumartojo, 2022).

Care, in this sense, is a phenomenon on the move, crossing conceptual, experiential, and material boundaries and, with that, subject to the same porosity that we have discussed in previous chapters (see also Atkinson et al., 2011, p. 569). In our understanding, then, care is simultaneously thought, practice, and feeling; it is rational, practical, and affective, and it harbours ethical imperatives and is therefore bound to relations centred on notions of responsibility and obligation. With that, care is embedded within both human relationships and the particular material environments that together make up the atmospheres in a city. It is also what gave Kim the sense that, in some areas at least, 'people look out for each other'.

Building care

Looking at the intersections between care and atmospheres in cities, which are also political spaces, further involves tending to the ways care might be lacking. As María Puig de la Bellacasa (2017, p. 1) has pointed out, 'care is omnipresent, even through the effects of its absence'. It is often fairly easily discerned, at least on a surface level, which parts of a city are cared for, and which parts have been subject to neglect over longer or shorter periods of time. However, care is not always immediately visible, nor does it take on any singular shape. It is about building and maintaining infrastructure, but it is also about facilitating particular experiences of what cities feel like (cf. Jaffe et al., 2020).

To make matters even more complicated, care is not an inherently benevolent phenomenon. It is, as Bellacasa writes, 'not only ontologically but politically ambivalent' (2017, p. 7). In other words, to care can also hurt, both the carer and the cared for. We care about people and places that we can do very little or nothing for. This reminds us of Danish philosopher K.E. Løgstrup (2020, p. 15; see also Jensen, 2018, p. 118), who once commented that:

> An individual never has something to do with another human being without holding something of that person's life in their hands. It can be a very small matter, a passing mood, a dampening or quickening of spirit, a disgust one deepens or takes away. But it may also be of tremendous significance, so that it is simply up to the individual whether the other person's life flourishes or not.

When someone encounters a homeless person, comes across a couple having a heated argument, or passes a building that has become dilapidated over years of neglect, they might be concerned but also feel powerless in their relative impotence. Sure, they can offer people an attentive look or offer assistance in the here and now, but people may not feel in a position to fundamentally change these lives and environments for the better. Apart from small interventions, the desire to care for the material environment is most often stifled by a lack of the right tools, permissions, and resources to improve them. Rather, this type of care is often in the Nordic countries carried out under the auspices of municipal and other authorities, leading us to the question of how cities are designed and built in relation to notions and experiences of care (see also Price et al., 2021; Rishbeth & Rogaly, 2018).

An example that highlights these complexities of how care works as an affective phenomenon is North West Park, the public park on the outskirts of Copenhagen that we have already visited several times throughout this book. The park was refurbished just over a decade ago and redesigned to be a playful, engaging, and fairy-tale-like public space that features colourful lighting, patterned light projections, and decorative elements such as striped lampposts and star-shaped concrete blocks scattered along the paths, which also showcase poems in the handwriting of local school-children (see Schwabe et al., submitted).

Cities of care 101

Figure 5.2 After dark at North West Park, Copenhagen.
Photo by Ida Lerche Klaaborg.

In the case of North West Park, while the park has been designed with careful attention to local particularities, users of the park do not necessarily all feel cared for or cared about in their experience of the park. Among the people we interviewed there, some indeed seemed to enjoy the design while many found it somewhat kitsch and a bit over-the-top. One group, meanwhile, had a different experience of the space altogether. Made up of local men and women, this loosely structured group came together on most days to drink, smoke, and enjoy each other's company in a nook of the park that was both perceived and designed as their designated area.

With benches and a small white tarp put in place to protect them from the worst of the rain on wet days, the area had been designed with input from this group of users, seemingly to encourage a sense of inclusion and even ownership of the space among them. It was a meeting place for them, somewhere to get together with old friends and new arrivals, and simultaneously to feel a sense of local attachment. However, one element was put in place without consultation with these regulars, namely what they referred to as the 'surveillance lamp': A bright, white spotlight, pointed right at the group, that would come on and switch off at regular intervals after dark while a green spotlight lit up the area more dimly. This bright light, we were told, was the cause of some frustration since visibility – in the sense of being able to see – would regularly shift between full visibility in the glare of the bright light and a more limited visibility as facilitated by the subdued green light. More interestingly, the workings of this 'surveillance lamp' was also the cause of visibility – in the sense of *being seen* – having become an affective focal point for the regulars, who noted that, while they might not be under surveillance in a strict sense, they felt as if they were.[1]

In this particular space, although the intention of the designers had been to create an open space that felt safe, they had also, with their design interventions, facilitated experiences of surveillance, of potentially being *watched* rather than merely *seen* (Schwabe et al., submitted). These experiences hinged on the atmosphere; apart from the artificial lights creating possibilities for someone to watch the regular users of the space, there were no overt surveillance technologies in place. In other words, the issue here was not that surveillance was indeed taking place, but rather that the design of the park made these users feel *hypervisible*, uneasy and on high alert, as if they *could* be watched. For the people that we spoke to in this area of the park, attuning to the space led them to feel as if their own presence was simultaneously welcome and precarious. Even when they spent time in their usual spot alone, they felt that they could potentially be watched and perhaps even asked to leave if someone were to consider them too noisy.

It should be noted that while it may have been the designers who made the decision to implement particular elements in the space – not least the 'surveillance lamp' – the redesign project had been initiated by the municipality while the usual upkeep of the park had long been taken over by municipal workers by the time of our research. This diffusion of power and responsibility is important to keep in mind as we continue to discuss how cities of care are not only built, but also maintained with more or less success by different people and institutions over time. Urban spaces that are perceived to harbour caring qualities upon construction require upkeep and ongoing caretaking to remain that way; conversely, the feel of spaces that may at first seem uncaring, even eerie (such as what Thomas and Anna described in Copenhagen and Oslo, respectively, in Chapter 2), can change over time to the extent they are subject to caring interventions. In North West Park, what may be considered a caring approach to urban design, where emphasis on the participation of locals played a role in the design phase, had nonetheless resulted in a divisive space where at least some users recurrently felt less than comfortable.

In line with our overall argument, we see care as necessarily as much about *affect* as *effect*. In other words, questions to do with how to build and maintain cities marked by care need responses that take into account care as a spatial *feeling* and not just the extent to which care as an activity is *effective* in urban environments. In the case of North West Park, it could be argued that those behind the redesign and maintenance of the park have carried out this work with care; they included users and residents in planning the refurbishment, added light and cut down shrubs that hindered visibility to secure a sense of safety, while aiming for a playful design that would encourage people to experience a sense of adventure and exploration in the space. But cities that feature a sense of care rely on more than systematised modes of caretaking of the built environment and measures that seek to make spaces effectively safe and easy to navigate. While such measures play an important role in facilitating experiences of care, we need to pay attention to their affective potential rather than only their effective results (cf. Hvass et al., 2022;

Sumartojo, 2022). The significance of this affective potential becomes clear in a case such as North West Park, where issues of visibility and surveillance *as feeling* rather than practical reality come to the fore.

Atmospheres of care

Kim's story, discussed in the beginning of this chapter, sheds further light on how care is experienced as an everyday phenomenon in the atmospheric city. Reflecting on past and present homes, Kim described how they grew up in a neighbourhood of small, detached houses where people would live their lives on their own 'side of the hedge', as they put it. In other words, there was little neighbourly interaction across property boundaries, and seemingly little sense of mutual care among the residents. In contrast, Kim explained, they now live in a social housing complex where the people around them are very different. In this setup, there is no escaping a certain sense of intimacy with the neighbours; an intimacy that is embodied and sensed in very direct ways. As Kim described:

> When I walk past the apartment right below mine, I can smell cigarette smoke in a way that I associate with my grandmother's and grandfather's home, where everything just smelled of smoke – but in a way I think it's quite cosy.

Explaining the cosiness of their neighbour's cigarette smoke, Kim went on:

> I am never in doubt as to whether they just left within the last half hour, because I can smell it in the stairwell. I think that's cosy because it's homely; the pleasant smell (*duft*) of people smoking cigarettes in the house.

The sense of intimacy and homeliness – and, with those, the comfort – that is brought on for Kim by the smell of their neighbour's cigarette smoke is also tied to other ways of relating to their surroundings. Next door to Kim lives a woman whose sister and her children often visit. As Kim said, she is very sweet and always very apologetic about the noise they make. Their presence is felt not so much via smells but via the noise of the kids playing, Kim explained to us: 'Those kids just have a great time and run around the apartment. It's just really cosy because I can always hear them chatting and stuff.' The importance of sound here does not eliminate the role of smell, however, as the two intermingle in Kim's experience. As they added, 'She always cooks super delicious food that you can smell all the way up the stairs… that always makes me really happy'.

During our conversation, Kim went on to describe to us the everyday routines and lives of their other neighbours; the retired Swedish professor downstairs who seems to live in her own world and gets startled when someone else happens to pass her on the stairs – something that, again, seems cosy to Kim, who likes to imagine what her world is like inside that apartment, full of books and listening to the radio – the next-door neighbour who is always late and so always in a rush, and the few people who have retired early. Age and (previous) profession are not the only markers of difference within the building. Just in the apartments accessed from Kim's stairwell, they explained, five or six different languages are spoken; 'we're just a very mixed group of people'. To Kim, all these neighbours are 'some people you meet and somehow you feel like you're a part of their daily life, but not quite like you are with your friends'. It seems that to Kim, the 'neighbor is not a stranger, nor a friend, nor kin but a form of sociality whose value is an effect of ambience-experimentation. The neighbor is an atmospheric person' (Jiménez & Estalella, 2013, p. 121). There is a soothing kind of comfort to knowing these atmospheric persons to Kim and, in turn, the reassurance of their vicinity make Kim feel that they are cared for (see also Bille 2019).

Kim's example shows how a sense of care is produced through the relationality and immediacy of atmospheres. In their building, and throughout the wider neighbourhood, Kim experiences care as what Alberto Corsín Jiménez and Adolfo Estalella call 'an atmospheric installation' (2013, p. 131) which is co-produced with their neighbours and perceived in subtle, sensory ways. It is important to note here that the comfort – reassurance and ease – and sense of mutual care that Kim experienced through atmospheres in their apartment building was as much facilitated by their material surroundings as by social circumstances. Their building is old and so insulation between apartments is lacking to the extent that both smells and sounds seem to travel rather freely between individual homes. However, while this may at the surface seem like a nuisance, as evidence of neglect, and a source of irritation – not being allowed the control of one's immediate surroundings that many people associate with a feeling of being at home – to Kim it was the very porosity of the sensory boundaries around their apartment that made them feel connected to their neighbours. In addition to that, the very atmosphere of homely comfort that Kim experienced there is, in a sense, porous. This atmosphere neither begins nor ends at the physical perimeter of their own private apartment but is something marked by malleable boundaries that extend beyond the home proper and which only slowly fades with distance.

In their apartment block, Kim finds themselves in a situation of constantly relating to their surroundings. Smells and sounds seep through walls, floors, and ceilings, and with them seep an awareness of the presence of others. At home, although protected by the physical boundaries of their apartment, Kim cannot help but know others and be known by them. The neighbours are all separated from each other but nevertheless mutually felt. Not only is there room for difference here, there is also mutual recognition between the residents. At the same time, as Kim pointed out, the relationships between neighbours cannot be compared to friendships; these are simply people who have become part of each other's everyday life (see also Jacobs, 1961). In a broad sense, Kim's surroundings at home resonate with them and give them a sense that the neighbours are 'in it together' as 'atmospheric persons' (Jiménez & Estalella, 2013). Following Rosa's (2019, p. 2020) thinking, the resonance Kim experiences as a comforting feeling of being at home among neighbours who smoke and make noise and cook delicious food and get startled in the stairwell is possible not least because of the balance struck here between intimacy and foreignness. At the same time, the experience Kim has of this environment being caring is also influenced by memories of past homes; the feeling is thus wrapped up in the sensorially imposing, existing knowledge of Kim's neighbours, and associations to previous experiences.

Building caring cities necessarily involves people who play active roles in shaping their surroundings by their very presence and their capability of taking care of themselves (Hasse, 2014a). To that we might add that care as an aspect of urban dwelling entails much more than simply urban dwellers taking it upon themselves to actively care for themselves; it also includes the many other facets of care that we have looked at here, not least receiving care, the materiality of care, and, importantly, cultivating an atmospheric sensation that (other) people care and are cared for.

Zones of comfort

With regard to the intricacies of what it means to live in cities full of – and/or lacking – care, two modes of engagement (both of which are present in Kim's and other participants' stories) strike us as particularly fruitful to consider: Comfort and compassion. While comfort denotes a sense of physical and emotional well-being – a state that comes about in relation to our surroundings – compassion invokes the notion of an outwardly oriented feeling; a feeling *for* someone or something.

How do people find comfort in particular places and within particular atmospheres, and what work of care goes into producing and maintaining a kind of comfort that is wrapped up in atmospheres and affective relations (see McNally et al., 2021)? In Copenhagen, we also spoke to Nanna, who lives part-time with her father in a small apartment. Among other things, Nanna told us about the feeling of relief she sometimes experiences upon going outside and leaving her home:

> There's more space to breathe outside on the street. And generally more room to be yourself and do what you want because [in the apartment] there's just one bedroom and not many places where you can be left alone (literally: *være i fred*, be at peace). So even though it's out in public, I sometimes think that there are more opportunities to be at peace (*i fred*) outside than inside.

The phrase *at være i fred*, to be at peace, carries a double connotation. While the literal translation of *fred* is peace, the phrase is most commonly used to describe the experience of being left alone, undisturbed. In Nanna's case, it is striking that peace is found outside the home rather than within its walls; that it is in the public spaces of the neighbourhood that she finds a sense of comforting solitude. To Nanna as well as Kim, being left alone among neighbours is to be left in peace while still feeling a sense of comfort in recognition and resonance. This sense is also largely what shapes how local neighbourhoods like these come to be lived as comfort zones and atmospheric spaces of care.

Of course, while both Kim's and Nanna's stories point to rather harmonious experiences of localised neighbourhood care, there is most often a continuous movement in people's experiences of comfort and discomfort, and one does not necessarily rule out the other in any given situation; people often find themselves with mixed feelings about a given situation, interaction, or place. In that sense, comfort and discomfort are not dichotomous categories but fluid; indeed, comfort is a process (Pickerill, 2015; Price et al., 2021). Whether a place gives us a sense of comfort or not depends on who we are, who we get to be in that place, and what is happening there at the given moment. In that sense, comfort is also a 'negotiated, bodily practice' (Price et al., 2021, p. 6).

While most Copenhageners we spoke to referred with great fondness to the Nørrebro neighbourhood, it also became clear during our research that it does not bring about feelings of ease and well-being among everyone all the time; quite the contrary. As touched on previously, Nørrebro occupies a fraught location in the imaginaries of most Danish people; it has been extensively covered in national – and sometimes international – media and has historically been a site of wide-reaching and sometimes violent contestations over space. In recent decades, Nørrebro has been a centre of gang-related clashes and an ongoing conflict between criminal groups making territorial claims. Indeed, the presence of gangs and the threat of violence they pose has long been an atmospheric marker of certain parts of the neighbourhood (see Bille & Jørgensen, 2022) but does not seem to deter a range of people, especially students and young families, from spending time in the neighbourhood and settling there, even as housing prices are skyrocketing. Especially to older non-residents, however, Nørrebro seems to some degree tarnished by this presence; it is simply too diverse, with *too many strangers* who do not seem to belong, even though they, of course, probably spend more time there than most, whether they live there or not.

One person who presented us with a perception of Nørrebro that was very different from Kim's was Jens, whose daughters had both settled in Nørrebro while he lived in a more affluent neighbourhood not too far away. Thinking back to a recent period of unrest in the area, Jens reflected: 'That wasn't fun. It was a strange feeling two years ago, when we had this long, long period, where the gang war was going on in Nørrebro.' Jens described the incessant noise of a helicopter hovering in the air above the city constantly, surveilling the movements below:

> It was there when we went to visit the girls, and there were visitation zones and police, and we heard [stories] about people – friends' kids – who lived in apartments with shootings happening right downstairs on the street and things like that.

To Jens, the experience of living through that period had changed his perception of the city. 'All of a sudden I started thinking, "can I drive down this street or should I take the detour?" It was surreal', he said, because this was the Copenhagen he knew so well. 'It was the same surroundings, the same people, but the rules had suddenly changed… and I knew that there were some participants in this society who played by different rules, without me being able to see who they were.' To Jens, it was 'strange to have these thoughts', especially because, as he put it, 'I don't think anyone was going to start shooting at me… but then again, sometimes they don't shoot so well.' To Jens, this was an experience of absolute discomfort spurred on by stories he had heard as well as sensory impressions, not least from the helicopter above making constant noise. But it was also a discomfort based on something less tangible, a feeling of not being entirely safe, of not being able to move freely along his usual routes in case he came upon a stray bullet.

While, to Kim, there was comfort and a sense of protection in the diversity of Nørrebro – allowing them to be just one out of many types of what might elsewhere be considered 'outsiders' – to Jens there was no sense that, *if* something were to happen to him here, he would be looked after. He was well beyond his comfort zone. As a corollary to that experience, the city – and his position in it – changed for him. The city became something foreign and strange, and he, in turn, considered how he might keep it at a distance.

Figure 5.3 Nørrebro, Copenhagen.
Photo by Siri Schwabe.

The juxtaposition of Kim's and Jens's experiences prompts us to ask what makes some people, things, places, and situations more compelling than others? How is it that only some capture people's attention and instigate in them a sense of responsibility and a willingness to respond? What makes people 'read a scene of distress not as a judgement against the distressed but as a claim on the spectator to become an ameliorative actor' (Berlant, 2004, p. 1)? With reference to the work of Susan Sontag, Lauren Berlant points out that compassion can be felt as a result of perceived impotence in a situation involving suffering; but, she adds, it can also be understood as 'the apex of affective agency among strangers' (2004, 9). To Kim, Nørrebro is a place of both comfort and compassion; they can relax in the sensed knowledge that compassion and mutual commitment are to be found here. The reverse is the case for Jens. While he enjoys spending time in the neighbourhood, although perhaps less during times of heightened tensions, he would not consider living there; rather, he appreciates what he experiences as the comfort of living in a neighbourhood that is neat, calm, and where people tend to share more similarities than differences with himself.

When people feel entirely comfortable in a place, they often tell us that it feels almost like an embrace from their environment. When the people, the environment, and the atmosphere – including the air and the weather – that surround someone all fall into place, that person is most likely to feel calm and have a pleasant sense of being present. Ideally, perhaps, everyone should feel that way most of the time. However, everyone rarely does. Living in a city also means being surrounded by unpredictability; in Nordic cities at least, the weather is unstable at best, the surroundings change with rush-hour traffic and other changes in tempos and movements, and there are almost always people around whose actions and behaviour can never entirely be anticipated or prepared for. For many urban dwellers and visitors, this even seems to be part of the attraction of city life. No doubt, it is part of what makes cities atmospheric and underscores how perceptions of care inform how such atmospheric cities are experienced.

Knowing the city, feeling for the city

To feel comfortable in a given space is often about knowing that space, of being familiar with it, and perhaps even feeling a special affinity to it, a sense of belonging. Echoing our concern with porosity, Ahmed writes: 'To be comfortable is to be so at ease with one's environment that it is hard to distinguish where one's body ends and the world begins' (2007, p. 158; cited in Price et al. 2021, p. 6). But there are different levels to both comfort and familiarity, and the two overlap in various ways; indeed, people might very well feel comfortable immediately upon arriving somewhere new, just as they would when settling into a familiar spot. That is, comfort does not necessarily require familiarity, although it might be intensified by it. So, what is it that makes us feel safe, calm, and cared for in and by certain spaces?

As intimated above, to be comfortable often involves feeling cared for; feeling enveloped in a caring environment in which one can relax. Again, care is as much a feeling – an affective engagement with the world – as it is a practice. In the language of the Danish Crime Prevention Council (Scherg, 2018, p. 3; our translation from Danish), the experience of a lack of safety is made up of a triad of a 'susceptible subject', an 'unsafety generator', and the 'absence of comforting conditions'. Safety in this perspective is seen as a

> ...mental condition of peace and absence of worry and anxiety. Safety in this perspective is not a particularly intense experience, but a more 'quiet' background feeling. In our part of the world, safety is, for most people, such a foundational part of our living conditions that we take it for granted. Safety is thus not particularly remarkable – we notice it only when it is not present. Our experience of safety is thus also characterised by being 'tacit' and a 'state of unworriedness'.

Care, like experiences of safety, is never entirely passive. Even when people are on the receiving end of care, when they are simply present and attuned to a pleasant atmosphere, feeling comfortable and enjoying the moment, they are still relating to their surroundings, still playing a role in constituting a given place as a social and affective environment. As we saw above, feelings of homeliness, care, and comfort are often wrapped up in sensory impressions stemming from factors in people's surroundings. At the same time, how someone enters a space as already attuned certainly influences to what extent they might perceive it as caring. In that regard, both particular imaginaries and lived memories play a role in shaping experiences of care.

In Oslo, Emma described how, as a newcomer, she had begun to feel comfortable in her new neighbourhood, Frogner. Emma was in her 20s when we spoke to her and had moved to Oslo about a year earlier after having lived in smaller towns in the region. At first, Emma said, she had felt uncomfortable in Frogner, which she had thought of as posh and usually financially inaccessible to people like herself. Indeed, she had not wanted to live there but was intrigued by the ad that led her to join her current flatmates in their shared apartment. Bringing her preconceived ideas into the city with her, Emma had felt wary about her surroundings and even looked out for elements in the neighbourhood that might confirm her prejudice, not least posh older ladies acting rudelys to shopkeepers. After settling in and exploring the neighbourhood on foot, however, Emma started to feel increasingly comfortable.

It was not just her immediate sensory impressions, but also associations with past experiences that attuned her to her new home in the city. Specifically, it was the memory of a board game. The child of German immigrant parents, Emma had strong memories from her childhood of playing the board game Heimlich & Co. in which players move colourful spy-like figures around the board, not revealing which colour belongs to which player until the very end. The architecture of Frogner, Emma explained, had reminded her of the town depicted on the board, full of old and quaint buildings with turrets and balconies. This memory had, with a bit of time, brought an intimacy to her relationship with the neighbourhood. Now it seemed much less strange and much more familiar. In a sense, it was with this intimacy that Emma ultimately began to forge a bond of care with Frogner as an atmospheric environment.

While comfort and care are not synonymous, they are tightly connected. Indeed, there are elements of comforting experiences such as the ones described in this chapter that point to ways in which care is experienced atmospherically. Just as safety is a state of 'unworriedness' that is often taken for granted, care is to a large degree something that is sensed through atmospheres that provide comfort: Ease and, relatedly, a sense of soothing reassurance that one's neighbourhood or other immediate surroundings offer precisely safety and mutual care among 'atmospheric persons'.

Caring for plurality

In Stockholm, we spoke to Elin, a student in her 20s, who described finding a sense of calm in what she calls an 'anonymous sphere'. Elin lived in Kungsholmen, a quiet residential area of the city, which she characterised as 'an environment that doesn't demand anything', and which she saw as distinctly unpretentious. This was unlike other parts of the city, which Elin perceived as being more demanding of people – demanding of status, a certain level of income, and a certain appearance. In Kungsholmen, Elin felt like there was room for someone like her. At the same time, she recognised this as her own personal experience rather than a reflection of a wider reality in which Kungsholmen truly would be for anyone and everyone. As she was quick to point out, even her own neighbourhood was expensive and not exactly poised to fix what she saw as a wide-reaching issue of segregation in metropolitan Stockholm (see Rokem & Vaughan, 2018). At the same time, there was *room for someone like her* and, with that, she felt at ease and at home. She felt comfortable and subject to care.

As with both Kim and Elin, the comfort of being undisturbed while feeling safe and recognised was tightly bound to a basic experience of resonance with their surroundings. In both cases, resonance is about a feeling of familiarity and involves the potential of mirroring oneself in one's surroundings, seeing oneself reflected in them, whether they bring forth comfort in a perception of relative sameness or in the feeling that there is space for difference.

If cities are 'a world of strangers' (Lofland, 1973; Simmel, 1971), discussions of care necessarily revolve around issues of recognition and questions of what and who is subject to comfort and compassion within particular spaces. In other words, what fits in and what seems out of place? Who belongs and who does not? Taking such questions seriously requires thinking about the significance of sameness and difference in building and living in caring cities. It also urges us to consider the role of compassion in forging ties across differences. In this section, we tackle the issue of care in relation to plurality, considering the spatial multiplicity of cities while delving further into how people inhabiting discrete bodies experience and co-constitute relations of urban care.

Cities are of course characterised by being relatively densely populated. Usually, people are surrounded by others when they dwell in or move through urban space. Most of the time, there are strangers around; people who live in flats have people next door, upstairs, and downstairs that they only know superficially, as atmospheric persons, or not at all, and when they leave their homes, they tend to encounter local residents, workers, and visitors to their neighbourhoods. Besides, human presences are only part of the bigger picture. As dealt with in earlier chapters, in cities, people are constantly surrounded by varying materialities; paved roads, cobblestoned alleyways, brick and concrete buildings, lampposts, countless street signs and billboards, and of course all the various means of transportation that play such a big role in how they take in and navigate their surroundings.

In a city like Copenhagen, recreational public spaces are never far away, and residents have fairly easy access to both woods and water at beaches, harbourfronts, and lakesides. In Frederiksberg Municipality, which forms a central part of the city of Copenhagen, the local government made it official policy in 2018 that everyone should be able to see a tree from their window. So, whether at home or out beyond the four walls of their houses, urban dwellers are persistently confronted with cities that are *busy*, heterogeneous spaces. Although they are still sometimes thought of as 'concrete jungles', contemporary cities, especially in the Nordic countries, are increasingly constituted by a much wider selection of materials, not all produced by humans, but all to some extent or other ordered by humans. As we discussed in Chapter 3, people find themselves constantly embraced by their environment, and in cities, this environment is inherently complex, made up of a plurality of presences.

Figure 5.4 Being together around the harbour bath at Islands Brygge, Copenhagen.
Photo by Mikkel Bille.

In her influential work on space, Doreen Massey (2005, p. 151) highlights what she calls the 'chance of space', which 'may set us down next to the unexpected neighbour'. Contingency, claims Massey, is characteristic of space as it is constituted through processes of both chaos and order. Put plainly, we can never entirely know or predict what a space will be as our environment is constantly unfolding before us. Importantly, and in line with the observations above, this contingency is predicated on what Massey calls a 'throwntogetherness' of the human and the non-human. In such a 'throwntogether' world, where the presences of human and non-human elements intersect at various points in the city, at times by design, often by chance, people are constantly having to navigate in relation to what and who is around them.

In throwntogether worlds full of strangers, attunement is not simply a process that allows people to find common ground and become aligned with one another. Indeed, it can also *create* strangers and spur on experiences of alienation (Ahmed, 2014, p. 13; Jaffe et al., 2020). Attunement, in such an understanding, is both about *being with* and about *not being with*. There are those (both human and non-human actors) that people share an atmospheric space with, and those with whom they are out of tune and thus not really *with*. In that regard, those who are somehow out of sync and thus not really *with* us become 'obstacles' who not only bring discomfort and frustration to a situation but also rattle the overall atmosphere (Ahmed, 2014, p. 20). Not uncommonly, 'strangers' come to be seen as harbingers of danger and therefore the cause of fear (Merry, 1981). Strangers, it follows, are often expressly uncared for; people may not care for them to be in their vicinity and may not care for them to be part of their neighbourhood. In other words, people do not tend to extend their compassion to strangers. In contemporary cities, however, we often find areas where difference is celebrated, and where it is precisely the experience of plurality that facilitates a sense of comfort, such as in Kim's case.

Nevertheless, some people are more easily and readily placed within the category of stranger. For instance, in the Nordic context, people of colour are often lumped together in the category of 'non-Western immigrant' in public discourse, no matter the details of individual biographies, and they have often been deemed 'more other than other others' (Ahmed, 2000; see Simonsen & Kofoed, 2020). This circumstance of course hinges on ideas of what constitutes the norm in a given place and at a given time, and which bodies and types of presences are thought of as acceptable and catered to within an environment.

At the same time, looking at this issue from a wider angle, we might find that the challenges of dwelling in and moving through the city in a body that in one way or another deviates from the norm will likely become apparent to most people at one point or another. Whether someone finds themselves trying to cross a busy street on crutches, navigate a narrow sidewalk with a pram or simply find it increasingly difficult to orient themselves in a space due to diminishing eyesight and failing memory, experiences of the city change as people go through changes. After suffering a stroke, Sennett (2018, p. 15) describes how he started perceiving and understanding his surroundings differently. He now had to 'make an effort', as he writes, both to get his body to stand and walk the way he wanted and to keep from becoming entirely disoriented when in crowded spaces. Interestingly, rather than narrowing his focus to his immediate surroundings, Sennet notes that his post-stroke condition urged him to expand his sense of his surroundings and that he, in that way, 'became attuned on a broader scale to the ambiguous or complex spaces through which I navigated' (2018, p. 16).

Looking at plurality in relations of care, we begin to notice how care is distributed among people and places. As we saw above, some people might experience some places as caring and others less so. Indeed, some places might work as caring environments only for some people as a result of their affective qualities and the atmospheres that come to mark them as felt spaces. What does all this mean for how people (are able to) resonate with their surroundings? Returning to the idea that resonance is a mode of relation (Rosa, 2020, p. 31) that involves a certain level of unpredictability and cannot be designed or engineered, how does the inherent plurality of cities affect not only how people live in them, but also how they experience care in and through them? How do embodied encounters with both human and non-human others and 'stranger others' (Ahmed, 2000) help shape caring or uncaring spaces? In Rosa's thinking, the 'other' in resonant relationships must remain to some extent foreign, something that 'eludes or resists us' (2020, p. 49). Resonance, in that view, thus relies on the element of surprise, on difference, and on the potential for mutual response and transformation in our meeting with the 'others' of our world (see also Sennett, 2017).

In a broad sense, then, if we follow this line of thinking, we might suggest that an atmospheric city marked by care is a city that facilitates experiences of resonance by allowing for a certain level of messiness and surprise; the kind that cannot easily be designed or planned for (see Harnack, 2018). Questions of inclusion and exclusion are central to this issue. If cities are becoming increasingly ordered spaces with entire neighbourhoods that seem to only provide room for presences that are deemed desirable from above – often taking the shape of clean streets, neatly designed parks, cookie-cutter buildings, and middle-class people – then we risk ending up with cities that, despite the apparent intentions of authorities and designers, are experientially uncaring (see also Schwabe, 2021). If care is about affect as much as effect, about feeling as much as practice, then approaching the city as a space of care involves investigation of the affective porosity between people, things, and places that give form to the atmospheric city as an inherently everchanging space.

The idea that care exists only in the space of intimate interhuman compassion seems outdated. Also in a broader perspective, the realms of care and politics are deeply embedded within one another (see Kavedžija, 2021; Nussbaum, 2013; Tronto, 2015). In her work on 'wounded cities', Karen Till (2012) presents what she calls a 'place-based ethics of care' as an antidote to politics that might injure urban spheres and the life within them. In Till's view, adopting place-based ethics of care entails that 'attending to, caring for, and being cared for by place and those that inhabit place are significant ethical and political practices that may work to constitute more democratic urban realms' (2012, p. 5). From such a perspective, care is a political project through which everyday practices move cities toward being more just and equitable.

Conclusion

In this chapter, we have shown how care is crucial not only to individual lives through intimate experiences of comfort and interhuman compassion but also to how cities and wider societies are lived, perceived, and materialised over time. As a starting point for including this perspective in our considerations of what the future of atmospheric cities might look like, we take care to be something loudly, but most of the time 'quietly political' (Askins, 2015). It is always entangled in relations of power, in the ongoing, everyday contestations over who and what belongs where, how, and why. Simultaneously, it is a mundane activity that is felt and that unfolds through material interactions.

What we have shown in this chapter is that care is as much about atmospheres and a constant process of attunement to the materiality of the city as it is about particular interhuman acts and notions of duty and obligation. It is both practice and ethics, but it is also something felt, a sense of people's place in the place they are in. In that sense, care is both a deeply social and material phenomenon. It is experienced with and through nonhuman beings, things, and places that carry atmospheric qualities. In short, care is as much about affect as effect; it is about how cities feel. Nudging us toward the concluding chapter of this book (Chapter 6), which looks toward the future of the atmospheric city, care is also about what happens next (cf. Ahmed, 2010). To *feel cared for* by and in cities, and to *care for* cities, is also to be committed to the future of life in urban environments, and perhaps to work towards what we have called life-oriented cities.

Note

1 For further considerations on light and surveillance, see Entwistle and Slater (2019); Otter (2008); Schivelbusch (1987); Brandon, et al. (2021).

Figure 6.1

6 The future of the atmospheric city

The COVID-19 pandemic had a remarkable impact on cities around the world. In the Nordic context and elsewhere, things like distance markers, hand sanitisers, test centres, and COVID-19 advice and warnings came to mark urban spaces, and so too did the absence of traffic and people. The pandemic and the advent of lockdowns also made city dwellers observe their surroundings in new ways, both in terms of noting the sensation of relative emptiness in certain areas, and of a hyper-awareness of what they touched, how close they were to other people, and to the potential danger of contagion lurking everywhere, permeating and connecting things and bodies. Themes such as what does it mean that a place is 'clean', what or who is 'dangerous', and what bodily tactics are engaged as people navigate cities came to the fore. In a wider sense, the pandemic called attention to how the atmospheric and sensuous aspects of urban life work as enablers and distractors in people's use of the city, including both public spaces and private homes.

While we have yet to see the long-term effects of this period on urban planning and design, the pandemic of the early 2020s underscored the importance of considering how cities feel as everyday spheres whose atmospheres have significant bearing on how they are lived. Already before the pandemic, however, the significance of urban atmospheres had become apparent for a very different reason. Indeed, the research presented in this book began from the observation of a trend whereby many urban spaces in Nordic cities are (re)designed for sensory effect, whether by adding adjustable colourful spotlights and illuminating trees for atmospheric effect or by planning spaces to reduce sensory impact, for instance, by creating environmental zones to battle air pollution and lower noise levels, or setting up lighting therapy installations in public places to combat Seasonal Affective Disorder in wintertime.

When looking at these emerging trends, there is an important distinction to be made between what is done for *atmospheric effect* to make a space 'inviting', 'adventurous', or 'hip', and design for *sensory effect* such as interventions geared toward lowering noise, air pollution, or reduce negative health impact (although such interventions of course also have atmospheric potential). The latter could be argued to have the notion of the 'human as biological body' at its foundation. The former approach, in contrast, is more scenographic and centres on shaping spaces for people as aesthetic beings, aimed at attuning

DOI: 10.4324/9781003379188-6

(perhaps even seducing) them to and by spaces. Our concern in this book has been to not merely see what politicians, media, or designers plan for, comment on, or aim for in their design, but also to see how people actually make sense of the atmospheric qualities of the city. This means addressing the everyday life of being alone or with others, of engaging urban environments, including the things and weather that make up the city, of perceiving and practising urban life through movement, and of experiencing care and comfort as they unfold in such contexts.

As evident in the stories we have told, making sense of the city is not something that can be reduced to ten different experiences, or even fully captured in words. Of course, there is a difference between being alone in a place or with friends; there are differences between walking and cycling. But the atmospheric effect, that of feeling a 'serene' city, an 'energy', or simply feeling 'free' or 'relaxed', may be similar across places and practices. Our point, though, is not that it would not potentially be possible to reduce all such atmospheric experiences to categories, for instance, from comfortable to uncomfortable, relaxed or hectic (cf. Schönhammer, 2018; Vogels, 2008; Westerink et al., 2008). What we, however, have wanted to show, is that the quality of such impressions offers insights into what it actually means when people say that 'bikes are freedom' or that they 'feel safe in Nørrebro'. 'Safety', for instance, is not merely about the visibility of one's surroundings, cleaning off graffiti, or avoiding potentially threatening groups of people. It is also a way of sensing that others care – regardless of the objective nature of such care.

In this book, we have thus sought to detail not so much what happens in popular – or even necessarily well-designed – spaces, but rather how it feels to be in cities, with all the complexities of dwelling within and moving between the nooks and crannies of urban worlds that contain a variation of new, old, and derelict spaces. These places are places where people meet their friends; where they enjoy a moment of quiet; where they come to feel unsafe and thus hurry through; where they feel other people's poverty; where they sit alone to mend their hearts; where they grow old and grow memories.

When we talk about the atmospheric city, we talk about a city marked by such ways of being and becoming attuned. Taking a close look at how atmospheres work as everyday phenomena leads to an understanding of *life* in the city as less about particular activities and more about what it means to be human within urban spheres, with all the affective responses and socio-material relations that then entails. This understanding has led us to call for a life-oriented approach to cities that considers the intricacies of people's circumstances and ways of living. This approach takes into account how lives play out in urban spheres and holds space for the experiential complexity of these lives. In doing this, we have also attended to the qualities of places that are not neatly curated, hoping that this may remind readers, planners, and designers of the need to also allow for more organic city development that cherishes spaces that resist curation and that harbour qualities that are not easily designed, and may not be for everyone.

To offer a last example of one such quality, we return to Robert, the kitchen worker living in the Northwestern part of Copenhagen. He had once lived in Nørrebro and felt how the middle class had increasingly taken over the area to a somewhat homogenising effect. The buildings were still the same, and so too were the sidewalks and street lamps. But the cars parked in the area had perhaps become slightly more expensive, the coffee shops a bit more 'hip'. No doubt real estate prices had skyrocketed, and while the area was still very diverse, the neighbours were also just not all the same as before. Having moved further out to the Northwest, Robert found that his new neighbourhood was still not so gentrified. Like many other participants, we asked him about his favourite place, and, after pausing to think, he came to what he seemed to think was a surprising conclusion. The place he liked the most, he said, was a small park – or actually a small strip of green space – which 'is really this un-ostentatious park, and there's nothing there. Just some gravel trail and a lake and a railway. But I just find it really neat in some way.'

What he was describing was a green urban space spanning a park area with a lake and an area more cramped in between an area of allotment gardens, a military area, a large road, and a railway. It had also been a hub of resistance activities during the Second World War, something Robert was aware of, and which added to his perception of the space. To Robert, it was a small oasis in the city. Rather than neatly designed with even rows of trees, flower beds, and nice, maintained gravel trails, it was just a green area and a slightly 'untamed park', more nature-like to him than other parks. Despite its location close to trains, Robert contended that 'still, it's all quiet and idyllic'. It may be a banal point that all spaces evoke some sort of atmosphere, but a lesson to take from the many examples in this book is to recognise that the quality of spaces is not solely sensed according to their design and popularity, but very much also according to other (sometimes unexpected) qualities, such as their capacity for providing a pocket of solitude, for allowing a sense of historical roots and grandeur, or to recognise the small wonders of small green areas cramped in between the infrastructural layers of a busy city.

On the other hand, of course, not everyone will share Robert's experience. Someone else might feel unsafe in the park or simply find it unattractive. Likewise, with relatively minor tweaks of its appearance, its atmosphere would likely shift rather starkly. For instance, if the benches there were covered in graffiti or destroyed, or there was a marked presence of people with hostile attitudes or undertaking criminal activities, this park would no doubt be marked as 'unsafe' to many. While many of the examples we have used addressed positive aspects of being in the city it is important not to fall for the glamorised picture in the media of Copenhagen (or other Nordic cities) as the 'coolest' city, or romanticising what is essentially in many places the territorial stigma associated with areas where lack of economic and social resources, political decisions and demographic imbalances have made their impact.

In our investigation of how various spaces in the three Nordic cities of this book feel, we have argued that there is a need to not only look at what they might feel like to individual subjects, but also to take notice of the intermingling of the atmospheric qualities of materiality, design, and social life in constituting such feelings. Atmospheres may appear deeply personal, but they are simultaneously collective, multi-sensuous, materially shaped, and form part of the formal properties of things at least as much as they come into being in relation to the people who feel them (Anderson, 2009). When considering how people sense atmospheres in general, it is tempting to make their experience fit into well-established categories of sight, hearing, or smell. That is, how do people experience the light, sound, or smell of/in a place? Yet, to understand the role of the human sensory apparatus in perceiving and being in atmospheres, attention must be paid to how people sense and make sense as whole beings, with all the nuances and complexities (cf. Howes, 2018; Rodaway, 1994; Urry, 2012).

Our approach has highlighted precisely the ways atmospheres are lived, perceived, and practised as everyday urban phenomena. With that, we have not sought to explore in detail wider political issues to do with urban development, nor have we delved very far into how atmospheres are shaped by the political dynamics of, for instance, gentrification, stigmatisation, or various forms of injustice manifested in cities. However, it should be noted that we often see atmospheric design employed in areas that are considered vulnerable or as harbouring social challenges. While many places in Europe tackle spaces and neighbourhoods with high crime rates and other challenges through 'dark design' and 'secured by design' or surveillance paradigms (Bille & Jørgensen, 2022; Davis, 1992; Ebbensgaard, 2020; Entwistle & Slater, 2019; Jensen, 2018; Sloane et al., 2016), we have shown how places with fraught reputations like North West Park, Blågårds Plads, Stovner, and others actually feature a type of atmospheric design, highlighting artistic interventions, particularly those involving lighting in public space while downplaying security and safety precautions. Rather than being instances of design for (a sense of) security in an overt sense, in these instances and others, design works in seductive ways, not least through the installation of atmospheric lighting and design, thus harking back to the notion of 'ambient power' (Allen, 2006; Bille & Jørgensen, 2022; Schwabe et al., submitted).

In other words, while traditional policing initiatives have centred on surveillance, patrolling, and bright security lighting, and elements of 'surveillance lighting' persist to some extent in the abovementioned squares, we also currently see an increase in design and lighting practices that seek to seduce and tackle problems through the staging of atmospheres to make places not just seen but felt, and to evoke a sense of community and responsibility (see also Welsh et al., 2021). Rather than simply spreading uniform lighting for visibility, the lighting paradigm that seems to have gained impact in Nordic countries is one that more systematically reflects upon lighting and other sensory tactics, offering more focus on experience than instrumentally improving visibility. And this trend seems only to increase in the future as technologies are developing, and the urban lighting industry is expanding to not only include engineers and technicians, but also designers, some of whom come from backgrounds in scenography.

At least two issues are at stake with trends in design interventions, which we find may become an issue in the future. One is how recognisable and homogenous the design interventions become when the common solution to solving social problems involves *more* lighting, cutting down trees, allowing for visibility, and, in particular, the use of scenographic spotlights that have become an increasingly common lighting solution in Nordic countries over the last decade. That is, the danger is that a 'one size fits all' approach is employed in socially deprived areas where scenographic lighting is installed – often including co-creation processes and other social initiatives – but in more or less the same way, generating an atmospheric geography whereby the design points out areas that are considered problematic. In other words, spaces marked by social challenges become recognisable within cities *exactly* because of the way they are neatly curated through design.

Secondly, there is the opposite danger that, in the effort to distinguish one place from another, the design interventions become more aimed at a competition among designers to come up with new ideas, technologies, and interventions that effectively risk creating a gap between local residents' sentiments and aesthetic preferences and design. That is, designs aimed at winning prizes – for being nice to *look at* – rather than necessarily becoming positive identity markers in local communities – as places to *be in*.

These questions are not unlike those raised in light of the increasing use of anti-terror design in urban spaces in the Nordic context and beyond. Urban spaces have seen several mass killings or terror attacks in recent years, ranging from stabbings to attacks using motorised vehicles. Design interventions, in turn, have sought to balance a fine line between militarising the urban using overtly security-oriented elements such as concrete blocks, gates, fences, and surveillance, and more aesthetic interventions such as concrete flower pots that restrict vehicular access to pedestrian areas while acting as beautifying elements in public space. A central point is, as Asher Ghertner et al., note (2020), that security is achieved through the staging of absence. That is, using technologies and things to obviate risks and negate threats. Employing such technologies, they argue, helps satisfy the sensory demands of security – of *feeling* that a particular place is safe (2020, p. 5). This, of course, does not necessarily entail security in any objective sense, where the danger is simply not at hand, but rather speaks to the sensation of a space feeling safe. One central question here in this move from instrumental protective strategy to one where such protection is dressed in aesthetic cloaks, is what the sensorial effect is of such different styles of defensive measurements; from them being sensorially upfront as 'security' to being relegated into the background of awareness as simply a statue or flower pot with a secondary purpose (cf. Coaffee, 2009; Dalgaard-Nielsen et al., 2016; Ilum, 2022; Low, 2017). Will relegating protection to the sensuous background, shape a more vulnerable population in the case of emergency?

Related to this way of using material objects in organising the city is the recent influx of so-called Smart Technologies into urban spaces, ranging from 'Data from the Sky' (a technology that allows for in-depth traffic monitoring), car number plate payment for parking or environmental zones, AI traffic lights regulation, to more encompassing projects like Google's Sidewalk Toronto as 'a micro-city outfitted with smart technologies that will use data to disrupt everything from traffic congestion to health care, housing, zoning regulations, and greenhouse-gas emissions'[1] and thus intimately tying together commercial companies' knowledge of urban spaces with governmental planning. This development of a so-called smart city, with new technologies and data collection methods, allows for new understandings of the contemporary city (cf. Augusto, 2021; Bezdecny & Archer, 2018; Madsen et al., 2022; Willis & Aurigi, 2020), although, as several scholars have pointed out, the very term 'smart city' may encapsulate widely different capacities, practices, and imaginaries (cf. Hollands, 2008; Kitchin, 2015; Wigley & Rose, 2020).

For the purposes of this book, we are not so concerned with the fraught issue of the increasing datafication of urban spaces that affords researchers, politicians, and planners with tools to understand and shape cities (cf. Dalsgaard et al., 2021; Gabrys, 2019; Mattern, 2021; Powell, 2021). The possibilities for gathering and using of data on everything from air, noise, and traffic is poised to influence wider narratives about spaces, not to mention how they are ultimately experienced. As Austin Zeiderman and Katherine Dawson rightly comment, 'as an all-purpose technological solution to social ills, the smart city is only the most recent in a long line of future visions that seek to make the world a better place, one city at a time' (2022, p. 269; see also Pinder, 2005). What concerns us regarding the future visions of smart city developments in a broad sense is the way technologies are used for atmospheric or sensory effect to ostensibly make for better cities. This tendency not only raises pertinent questions regarding how cities feel – and how they are designed to feel – now and in the future, but also points to the centrality of considering the ethical issues of planning, designing, and using technologies to foster urban atmospheres. Where to draw the line between technological interventions working in the background to increase well-being and health, and those that aim at social behaviour? If we accept the sort of 'bio-hacking' that is implied by installing light therapy in public spaces to battle SAD, for instance, then what about more invasive anti-loitering technologies as reported by Charlotte Walsh (2008) like 'zit bulbs' that highlight acne in young people's skin through pinkish light, or 'The Mosquito' loudspeaker that emits a high pitched frequency tone mostly audible to young people? Put simply, how might such technologies that work to affect the body and perceptions of it, shape cities as lived?

The latter interventions are of course more invasive, and potentially discriminatory than designing a place that attracts certain users by staging a social scene through design or reducing the health risks of living in cities of noise and pollution. Nonetheless, they bring about atmospheric effects and play into notions of what a good city is (cf. Amin, 2006). These atmospheric effects are not necessarily noticeable, but simply work through a subtle 'atmospherological' or 'ambient' power (Allen, 2006; Hasse, 2014, p. 223). The case of Liquid Light in Oslo, 'designed to dynamically adapt and follow each person as he or she moves around, which creates a sense of well-being and improves our perception',[2] is a case in point. This and the others mentioned above are technologies that not only shape an atmosphere, but that are also based on particular notions of well-being.

All of this raises questions about the ability, indeed imperative, for both users, planners, designers, and politicians to be attentive to atmospheres and their seductive potential (Biehl-Missal & Saren, 2012; Böhme, 2006, pp. 43–53; Wolf & Julmi, 2020). Even further, the parallel emergence of scenographic city design and data-oriented urban planning calls for a more focused attention on the people who feel, sense, and make sense of atmospheric cities. Often, as we hope it has become clear, how cities *feel* tends to be markedly different from what is commonly presented via quantitative data and what designers and planners have come to expect.

Notes

1 https://www.theatlantic.com/technology/archive/2018/11/google-sidewalk-labs/575551/ Accessed 20 August 2022.
2 https://afry.com/en/liquid-lightr. Accessed 19 November 2021.

Bibliography

Adey, P., Brayer, L., Masson, D., Murphy, P., Simpson, P., & Tixier, N. (2013). 'Pour votre tranquillité': Ambiance, atmosphere, and surveillance. *Geoforum, 49*, 299–309.
Ahmed, S. (2000). *Strange encounters: Embodied others in post-coloniality*. New York: Routledge.
Ahmed, S. (2007). A phenomenology of whiteness. *Feminist Theory, 8*(2), 149–168. https://doi.org/10.1177/1464700107078139
Ahmed, S. (2010). *The promise of happiness*. Durham, NC. Duke UP.
Ahmed, S. (2014). Not in the mood. *New Formations: A Journal of Culture/Theory/Politics, 82*, 13–28.
Albertsen, N. (2019). Urban atmospheres. *Ambiances, 24*, 0–21.
Aldred, R. (2010). 'On the outside': Constructing cycling citizenship. *Social & Cultural Geography, 11*(1), 35–52.
Allen, J. (2006). Ambient power: Berlin's Potsdamer Platz and the seductive logic of public spaces. *Urban Studies, 43*(2 SPEC. ISS.), 441–455.
Amin, A. (2006). The good city. *Urban Studies, 43*, 1009–1023.
Amin, A., & Thrift, N. (2002). *Cities: Reimagining the urban*. Cambridge: Polity Press.
Anderson, B. (2009). Affective atmospheres. *Emotion, Space and Society, 2*, 77–81.
Anderson, B., & Ash, J. (2015). Atmospheric methods. In P. Vannini (Ed.), *Non-representational methodologies* (pp. 34–51). New York: Routledge.
Andersson, S. L. (2014). *The empowerment of aesthetics. Catalogue for the Danish Pavilion at the 14th International Architecture Exhibition La Biennale di Venezia*. Skive: Wunderbuch.
Ash, J., & Gallacher, L. A. (2015). Becoming attuned objects, affects, and embodied methodology. *Methodologies of Embodiment: Inscribing Bodies in Qualitative Research, 2014*, 69–85.
Askins, K. (2015). Being together: Everyday geographies and the quiet politics of belonging. *ACME 14* (2), 470–478.
Atkinson, S., Lawson, V., & Wiles, J. (2011). Care of the body: Spaces of practice. *Social and Cultural Geography, 12*, 563–572. https://doi.org/10.1080/14649365.2011.601238
Augusto, J. C. (Ed.). (2021). *Handbook of smart cities*. Cham: Springer.
Bakshi, A. (2017). *Topographies of memory: A new poetics of commemoration*. Cham: Palgrave Macmillan.
Beaumont, M. (2015). *Nightwalking. A nocturnal history of London*. London: Verso.
Bellacasa, M. P. d. l. (2017). *Matters of care. Speculative ethics in more than human worlds*. Minnesota, MN: University of Minnesota Press.

Benjamin, W., & Lacis, A. (1925). "Naples," in P. Demetz (Ed.), *Reflections: Essays, aphorisms, autobiographical writings* (pp. 163–173). New York: Bloch, 1978.
Berlant, L. (2004). Introduction: Compassion (and withholding). In Lauren Berlin (Ed.), *Compassion: The culture and politics of an emotion*. New York: Routledge.
Bezdecny, K., & Archer, K. (Eds.). (2018). *Handbook of emerging 21st-century cities.* Cheltenham: Edward Elgar Publishing Limited.
Biehl-Missal, B., & Saren, M. (2012). Atmospheres of seduction: A critique of aesthetic marketing practices. *Journal of Macromarketing, 32*, 168–180.
Bille, M. (2015). Lighting up cosy atmospheres in Denmark. *Emotion, Space and Society, 15*, 56–63.
Bille, M. (2017). Ecstatic things: The power of light in shaping Bedouin homes. *Home Cultures, 14*, 25–49.
Bille, M. (2019). *Homely atmospheres and lighting technologies in Denmark. Living with Light*. London: Bloomsbury.
Bille, M., & Hauge, B. (2022). Choreographing atmospheres in Copenhagen: Processes and positions between home and public. *Urban Studies, 59*(10), 2076–2091.
Bille, M., & Jørgensen, O. N. (2022). At the margins of attention: Security lighting and luminous art interventions in Copenhagen. In S. Sumartojo (Ed.), *Lighting Design in Shared Public Spaces*. New York: Routledge, s. 125–150 26 s.
Bille, M., Bjerregaard, P., & Sørensen, T. F. F. (2015). Staging atmospheres. Materiality, culture and the texture of the in-between. *Emotion, Space & Society, 15*, 31–38.
Bille, M., & Simonsen, K. (2021). Atmospheric practices: On affecting and being affected. *Space and Culture, 24*(2), 295–309.
Binswanger, L. (1933). Das Raumproblem in der Psychopathologie. *Zeitschrift Für Die Gesamte Neurologie Und Psychiatrie, 145*, 598–647.
Bissell, D. (2008). Comfortable bodies: Sedentary affects. *Environment and Planning A, 40*(7), 1697–1712.
Bissell, D. (2009). Conceptualising differently-mobile passengers: Geographies of everyday encumbrance in the railway station. *Social and Cultural Geography, 10*(2), 173–195.
Bissell, D. (2010). Passenger mobilities: Affective atmospheres and the sociality of public transport. *Environment and Planning D: Society and Space, 28*(2), 270–289.
Bissell, D. (2018). *Transit life: How commuting is transforming our cities*. Cambridge, MA: MIT Press.
Bissell, D., & Fuller, G. (Eds.). (2011). *Stillness in a mobile world*. New York: Routledge.
Blanco, H. (2018). Liveable cities: From concept to global experience. In R. Caves, & F. Wagner (Eds.), *Liveable cities from a global perspective* (pp. 1–13). New York: Routledge.
Böhme, G. (1993). Atmosphere as the fundamental concept of a new aesthetics. *Thesis Eleven, 36*, 113–126.
Böhme, G. (1995). *Atmosphäre: Essays zur neuen Ästhetik*. Frankfurt am Main: Suhrkamp.
Böhme, G. (1998). Atmosphere as an aesthetic concept. *Daidalos, 68*, 112–115.
Böhme, G. (2001). *Aisthetik. Vorlesungen über Ästhetik als allgemeine Wahrnemungslehre.* München: Wilhelm Fink.
Böhme, G. (2006). *Architektur und Atmosphäre*. München: Wilhelm Fink GmbH.
Böhme, G. (2011). Das Wetter und die Gefühle. Für eine Phänomenologie des Wetters. In K. Andermann & U. Eberlein (Eds.), *Gefühle als Atmosphären. Neue Phänomenologie und philosophische Emotionstheorie* (pp. 153–166). Berlin. Akademie Verlag.
Böhme, G. (2017). *The aesthetics of atmospheres*. New York: Routledge.

Bibliography

Bollnow, O. F. (1941). *Das Wesen der Stimmungen*. Frankfurt am Main: Klostermann.

Bollnow, O. F. (1963). *Mensch und Raum*. Stuttgart: Kohlhammer.

Borch, C. (2013). Crowd theory and the management of crowds: A controversial relationship. *Current Sociology*, *61*(5–6), 584–601.

Brennan, T. (2004). *The transmission of affect*. Ithaca, NY: Cornell University Press.

Brighenti, A. M., & Pavoni, A. (2019). City of unpleasant feelings. Stress, comfort and animosity in urban life. *Social and Cultural Geography*, *20*(2), 137–156.

Bull, M. (2007). *Sound moves: iPod culture and Urban experience*. London: Routledge.

Buttimer, A. (1976). Grasping the dynamism of lifeworld. *Annals of the Association of American Geographers*, *66*, 277–292.

Casey, E. (2022). *Turning emotion inside out*. Evanston, IL: Northwestern University Press.

Catucci, S., & De Matteis, F. (Eds.). (2021). *The affective city—spaces, atmospheres and practices in changing urban territories*. Sericuse: Lettera Ventidue.

Ceccato, V. (2020). The architecture of crime and fear of crime: Research evidence on lighting, CCTV and CPTED features. In V. Ceccato & M. K. Nalla (Eds.), *Crime and fear in public places. Towards safe, inclusive and sustainable cities* (pp. 38–72). London: Routledge.

Chowdhury, R., & McFarlane, C. (2021). The crowd and citylife: Materiality, negotiation and inclusivity at Tokyo's train stations. *Urban Studies*, *59*, 1353–1371.

Coaffee, J. (2009). Protecting the urban: The dangers of planning for terrorism. *Theory, Culture & Society*, *26*(7–8), 343–355.

Cobe. (2018). *Our urban living room*. Stockholm: Arvinius + Orfeus Publishing.

Cole, T. (2011). *Open city*. New York: Random House.

Conradson, D. (2003). Geographies of care: Spaces, practices, experiences. *Social & Cultural Geography*, *4*(4), 451–454.

Corbin, A. (2014). Urban sensations: The shifting sensescape of the city. In C. Classen (Eds.), *A cultural history of the senses in the age of empire* (pp. 47–68). London: Bloomsbury Academic.

Coverley, M. (2018). *Psychogeography*. Harpenden England. Oldcastle Books.

Cresswell, T. (2004). *Place: A short introduction*. London: Blackwell.

Criado Perez, C. (2020). *Invisible women: Exposing data bias in a world designed for men*. London: Vintage.

Dalgaard-Nielsen, A., Laisen, J., & Wandorf, C. (2016). Visible counterterrorism easures in urban spaces—fear-inducing or not?" *Terrorism and Political Violence*, *28*(4), 692–712.

Dalsgaard, S., Haarløv, R. T., & Bille, M. (2021). Data witnessing: Making sense of urban air in Copenhagen, Denmark. *HAU: Journal of Ethnographic Theory*, *11*(2), 521–536 16 s.

Davidson, J., Bondi, L., & Smith, M. (Eds.). (2005). *Emotional geography*. Aldershot: Ashgate.

Daniels, I. (2015). Feeling at home in contemporary Japan: Space, atmosphere and intimacy. *Emotion, Space and Society*, *15*, 47–55.

Davis, J. (2022). *The caring city: Ethics of urban design*. Bristol: Bristol University Press.

Davis, M. (1992). *City of Quartz. Excavating the future in Los Angeles*. New York: Vintage Books.

De Matteis, F. (2018). The city as a mode of perception: Corporeal dynamics in urban space. In F. Aletta & J. Xiao (Eds.), *Handbook of research on perception-driven approaches to Urban assessment and design* (pp. 436–457). Hershey PA: IGI Global.

De Matteis, F. (2020). *Affective spaces. Architecture and the living body*. London: Routledge.

Degen, M., & Lewis, C. (2019). The changing feel of place: The temporal modalities of atmospheres in Smithfield Market, London. *Cultural Geographies, 27*(4), 509–526.

Degen, M., & Rose, G. (2022). *New urban aesthetics. Digital experiences of urban change*. London. Bloomsbury.

Degen, M., Melhuish, C., & Rose, G. (2017). Producing place atmospheres digitally: Architecture, digital visualisation practices and the experience economy. *Journal of Consumer Culture, 17*(1), 3–24.

DeSilvey, C. (2017). *Curated decay: Heritage beyond saving*. Minneapolis, MN: University of Minnesota Press.

Donald, J. (1999). *Imagining the modern city*. London: Athlone.

Dreyfus, H. L. (2012). Why the mood in a room and the mood of a room should be important to architects. In B. Jaquet, & V. Giraud (Eds.), *From the things themselves: Architecture and phenomenology*. Kyoto: Kyoto University Press.

Duff, C. (2010). On the role of affect and practice in the production of place. *Environment and Planning D: Society and Space, 28*(5), 881–895.

Dunn, N., & Edensor, T. (Eds.). (2022). *Rethinking darkness: Cultures, histories, practices*. London: Routledge.

Durkheim, E. (1912 [1995]). *The elementary forms of religious life*. New York: The Free Press.

Durkheim, E. (1982). *The rules of sociological method*. New York: The Free Press.

Ebbensgaard, C. L. (2020). Standardised difference: Challenging uniform lighting through standards and regulation. *Urban Studies, 57*, 1957–1976.

Edensor, T. (2005). *Industrial ruins: Space, aesthetics and materiality*. London: Berg.

Edensor, T. (2015). Producing atmospheres at the match: Fan cultures, commercialisation and mood management in English football. *Emotion, Space and Society, 15*, 82–89.

Edensor, T. (2017). *From light to dark: Daylight, illumination and gloom*. Minneapolis, MN/London: University of Minnesota Press.

Edensor, T., & Sumartojo, S. (2015). Designing atmospheres: Introduction to special issue. *Visual Communication, 14*(3), 251–265.

Eichler, M. (Ed.). (1995). *Change of plans: Towards a non-sexist sustainable city*. Toronto: Garamond Press.

Entwistle, J., & Slater, D. (2019). Making space for 'the social': Connecting sociology and professional practices in urban lighting design1. *British Journal of Sociology, 70*, 1–22.

Ellard, C. (2015). *Places of the heart: The psychogeography of everyday life*. New York. Bellevue Literary Press.

Ferran, I. (2022). Moods and atmospheres: Affective states, affective properties, and the similarity explanation. In D. Trigg (Ed.), *Atmospheres and shared emotions* (pp. 57–74). Routledge.

Fisher, B., & Tronto, J. C. (1990). Toward a feminist theory of caring. In Emily K. Abel, & Margaret Nelson (Eds.), *Circles of care* (pp. 36–54). Albany, NY: SUNY Press.

Fisher, M. (2016). *The weird and the eerie*. London: Repeater Books.

Fitz, A., & Krasny, E. (2019). *Critical care: Architecture and urbanism for a broken planet*. Cambridge, MA: MIT Press.

Frølund, S. (2018). Gernot Böhme's sketch for a weather phenomenology. *Danish Yearbook of Philosophy*, *51*(1), 142–161.

Fujii, J. A. (1999). Intimate alienation: Japanese urban rail and the commodification of urban subjects. *Differences*, *11*(2), 106–133.

Gabrys, J. (2019). *How to do things with sensors*. Minneapolis, MN: University of Minnesota Press, Forerunners series.

Gandy, M. (2017). Urban atmospheres. *Cultural Geographies*, *24*(3), 353–374.

Gehl, J. (2010). *Cities for people*. London: Island Press.

Gehl, J. (2011). *Life between buildings: Using public space*. London: Island Press.

Gehl, J., & Svarre, B. (2013). *How to study public life*. London: Island Press.

Gherardi, S. (2017). One turn ... and now another one: Do the turn to practice and the turn to affect have something in common? *Management Learning*, *48*(3), 345–358.

Ghertner, A. D., McFann, H., & Goldstein, D. M. (Eds.). (2020). *Futureproof: Security aesthetics and the management of life*, Durham, NC: Duke University Press.

Giacomin, J. (2014). What is human centred design? *Design Journal*, *17*(4), 606–623.

Gieryn, T. F. (2002). What buildings do. *Theory and Society*, *31*(1), 35–74.

Griffero, T. (2014). *Atmospheres: Aesthetics of emotional spaces*. Aldershot: Ashgate.

Griffero, T. (2021). *The atmospheric "we". Moods and collective feelings*. Mimesis International.

Griffero, T., & Tedeschini, M. (Eds.) (2019). *Atmosphere and aesthetics. A plural perspective*. Basingstoke: Palgrave.

Groening, S. (2014). *Cinema beyond territory: Inflight entertainment and atmospheres of globalization*. Basingstoke: Palgrave Macmillan.

Halbwachs, M. (1980). *The collective memory*. New York: Harper Colophon.

Hanich, J. (2021). Shared or spread? On boredom and other unintended collective emotions in the cinema. In D. Trigg (Ed.), *Atmospheres and shared emotions* (pp. 135–151). London: Routledge.

Harnack, M. (2018). Drifting clouds: Porosity as a paradigm. In S. Wolfrum, H. Stengel, F. Kurbasik et al. (Eds.), *Urban porosity and the right to a shared city. Porous city* (pp. 38–41). Basel: Birkhaüser.

Hasse, J. (2010). Atmosphären und Stimmungen im Denmalschutz—Zur Überwindung des Visualismus im Denkmalschutz. *Die Denkmalpflege*, *68*(2), 108–126.

Hasse, J. (2012). *Atmosphären der Stadt. Aufgespürte Räume*. Berlin: Jovis.

Hasse, J. (2014a). Atmospheres as expression of medial power. Understanding atmospheres in urban governance and under self-guidance. *Lebenswelt. Aesthetics and Philosophy of Experience*, *4*(1), 214–229.

Hasse, J. (2014b). *Was Räume mit uns machen—und wir mit ihnen*. München: Verlag Karl Alber.

Hasse, J. (2018). *Märkte und ihre Atmosphären. Mikrologien räumlichen Erlebens*. München: Verlag Karl Alber.

Hasse, J. (2019). Atmospheres and moods: Two modes of being-with Jürgen. In T. Griffero, & M. Tedeschini (Eds.), *Atmosphere and Aesthetics*. Basingstoke: Palgrave.

Hebbert, M. (2005). The street as locus of collective memory. *Environment and Planning D: Society and Space*, *23*, 581–596.

Heidegger, M. (1995). *The fundamental concepts of metaphysics: World, finitude, solitude*. Bloomington: Indiana University Press.

Heidegger, M. (1996). *Being and time*. New York: State University of New York Press.

Hollands, R. G. (2008). Will the real smart city please stand up? Intelligent, progressive or entrepreneurial? *City, 12*(3), 303–320.

Hommels, A. (2005). *Unbuilding cities: Obduracy in urban socio-technical change.* Cambridge, MA: MIT Press.

Houser, K. W., Boyce, P. R., Zeitzer, J. M., & Herf, M. (2021). Human-centric lighting: Myth, magic or metaphor? *Lighting Research and Technology, 53*(2), 97–118.

Howes, D. (2018). Introduction. Make it new. Reforming the sensory world. In D. Howes (Ed.), *A cultural history of the senses in the modern age.* London: Bloomsbury.

Howes, D. (2019). Multisensory anthropology. *Annual Review of Anthropology, 48,* 17–28.

Howes, D., & Classen, C. (2014). *Ways of sensing: Understanding the senses in society.* London and New York: Routledge.

Hvass, M., et al. (2022). Lights out? Lowering urban lighting levesle and increasing atmosphere at a Danish tram station. In S. Sumartojo (Ed.), *Lighting design in shared public spaces* (pp. 151–172). London: Routledge.

Ifversen, K. R. S. (2019). The long haul. In K. L. Weiss (Ed.), *Critical city* (pp. 131–139). Arkitektens Forlag: København.

Ilum, S. (2022). *A significant threat. Countering terrorism in the good city.* PhD Thesis. Copenhagen, University of Copenhagen.

Ingold, T. (2000). *The perception of the environment: Essays in livelihood, dwelling and skill.* London: Routledge.

Ingold, T. (2007). Earth, sky, wind, and weather. *Journal of the Royal Anthropological Institute, 13,* S19–S38.

Ingold, T. (2011). *Being alive. Essays on movement, knowledge and description.* London: Routledge.

Ingold, T. (2015). *The life of lines.* London: Routledge.

Ingold, T., & Vergunst, J. L. (2008). Introduction. In T. Ingold, & J. L. Vergunst (Eds.), *Ways of walking. Ethnography and practice on foot* (pp. 1–20). London: Routledge.

Jacobs, J. (1961). *The death and life of great american cities.* New York: Random House.

Jaffe, R., Dürr, E., Jones, G. A., Angelini, A., Osbourne, A., & Vodopivec, B. (2020). What does poverty feel like? Urban inequality and the politics of sensation. *Urban Studies, 57*(5), 1015–1031.

Jensen, O. B. (2018). Dark design: Mobility injustice materialized. In Nancy Cook, & David Butz (Eds.), *Mobilities, mobility justice and social justice* (pp. 116–128). London: Routledge.

Jensen, O. B. (2020). Atmospheres of rejection: How dark design rejects homeless in the city. *Ambiences, Alloæsthesia: Senses, Inventions, Worlds: Proceedings of the 4th International Congress on Ambiances,* 326–331.

Jensen, O. B. (2022). Urban mobilities and power: Social exclusion by design in the city. In Nadine Cattan, & Laurent Faret (Eds.), *Hybrid mobilities: Transgressive spatialities* (pp. 37–55). London: Routledge.

Jensen, O. B., Sheller, M., & Wind, S. (2015). Together and apart: Affective ambiences and negotiation in families' everyday life and mobility, *Mobilities, 10*(3), 363–382.

Jensen, O. B., Martin, M., & Löchtefeld, M. (2021, September). Pedestrians as floating life—on the reinvention of the pedestrian city. *Emotion, Space and Society, 41,* 100846.

Jiménez, A. C., & Estalella, A. (2013). The atmospheric person: Value, experiment, and "making neighbors" in Madrid's popular assemblies. *HAU: Journal of Ethnographic Theory, 3,* 119–139.

Bibliography

Jones, O., & Garde-Hansen, J. (Eds.). (2012). *Geography and memory: Explorations in identity, place and becoming*. Basingstoke: Palgrave Macmillan.

Kaal, H. (2011). A conceptual history of liveability. *City, 15*, 532–547.

Kavedžija, I. (2021). *The process of Wellbeing: Conviviality, care, creativity (Elements in psychology and culture)*. Cambridge, MA: Cambridge University Press.

Kazig, R., Masson, D., & Thomas, R. (2017). Atmospheres and mobility. An introduction. *Mobile Culture Studies. The Journal, 3*(3), 7–20.

Kern, L. (2021). *Feminist city: Claiming space in a man-made world*. London: Verso.

Kitchin, R. (2015, March). Making sense of smart cities: Addressing present shortcomings, *Cambridge Journal of Regions, Economy and Society, 8*(1), 131–136.

Klingenberg, E. (2018). *Palaces for the people: How social infrastructure can help fight inequality, polarization, and the decline of civic life*. London: Penguin.

Klingenberg, E. (2002). *Heatwave. A social autopsy of disaster in Chicago*. Chicago, IL: University of Chicago Press.

Klingman, A. (2007). *Brandscapes. Architecture in the experience economy*. Cambridge, MA: MIT Press.

Knabb, K. (Ed.). (1995). *Situationist international anthology*. Berkeley, CA: Bureau of Public Secrets.

Koenig, J. (2021). *Dictionary of obscure sorrows*. New York: Simon & Schuster.

Kotler, P. (1974). Atmospherics as a marketing tool. *Journal of Retailing, 49*, 48–64.

Laage, G. (Ed.). (2005). *Die emotionale Stadt. Vom Planen, Bauen,und den Gefühlen der Bewohner*. München: Dölling u. Galitz.

Larkin, B. (2013). The politics and poetics of infrastructure. *Annual Review of Anthropology, 42*, 327–347.

Larsen, J. (2017). The making of a pro-cycling city: Social practices and bicycle mobilities. *Environment and Planning A, 49*(4), 876–892.

Larsen, J. (2022). *Urban marathon. Rhythms, places, mobilities*. London: Routledge.

Latham, A., & Layton, J. (2019). Social infrastructure and the public life of cities: Studying urban sociality and public spaces. *Geography Compass, 13*(7), 1–15.

Latham, A., & McCormack, D. P. (2017). Affective cities. In Michael Silk, David Andrews, & Holly Thorpe (Eds.), *Routledge Handbook of Physical Cultural Studies* (pp. 369–377). New York: Routledge.

Laurier, E., Lorimer, H., Brown, B., Jones, O., Juhlin, O., Noble, A., Perry, M., Pica, D., Sormani, P., Strebel, I., Swan, L., Taylor, A., Watts, L., & Weilenmann, A. (2008). Driving and passengering: Notes on the ordinary organisation of car travel. *Mobilities, 3*, 1–23.

Lawson, V. (2007). Geographies of care and responsibility. *Annals of the Association of American Geographers, 97*, 1–11.

Leder, D. (1990). *The absent body*. Chicago, IL: Chicago University Press.

Lefebvre, H. (2004). *Rhythm analysis: Space, time and everyday life*. London and New York: Continuum Books.

Lin, W. (2021). Assembling a great way to fly: Performances of comfort in the air. In Danny McNally, Laura Price, Philip Crang (Eds.), *Geographies of Comfort* (pp. 151–170). London: Routledge.

Löfgren, O. (2014). Urban atmospheres as brandscapes and lived experiences. *Place Branding and Public Diplomacy, 10*(4), 255–266.

Löfgren, O. (2015). Sharing an atmosphere: Spaces in urban commons. In C. Borch & M. Kornberger (Eds.), *Urban commons: Rethinking the city* (pp. 68–91). New York: Routledge.

Lofland, J. (1982). Crowd joys. *Urban Life, 10*(4), 355–381.

Lofland, L. H. (1973). *A world of strangers: Order and action in urban public space*. New York: Basic Books.

Løgstrup, K. E. (2020). *The ethical demand*. Oxford. Oxford University Press.

Low, S. M. (2017). Security at home: How private securitization practices increase. State and capitalist control. *Anthropological Theory, 17*(3), 365–385.

Madsbjerg, C. (2017). *Sensemaking: What makes human intelligence essential in the age of the algorithm*. London: Little, Brown Book Group.

Madsbjerg, C., & Rasmussen, M. B. (2014). *The moment of clarity: Using the human sciences to solve your toughest business problems*. Boston, MA: Harvard Business Review Press.

Madsen, A. K., Grundtvig, A., & Thorsen, S. (2022). Soft city sensing: A turn to computational humanities in data-driven urbanism, *Cities, 126*, 103671.

Mah, A. (2012). *Industrial ruination, community, and place: Landscapes and legacies of urban decline*. Toronto: University of Toronto Press.

Malpas, J. E. (1999). *Place and experience: A philosophical topography*. Cambridge, MA: Cambridge University Press.

Manning, E. (2013). *Always more than one*. Durham: Duke University Press.

Marinucci, L. (2019). Mood, ki, humors: Elements and atmospheres between Europe and Japan. *Studi di Estetica, 14*, 169–192.

Massey, D. (1994). *Space, place and gender*. Cambridge, MA: Polity Press.

Massey, D. (2005). *For space*. London: Sage.

Mattern, S. (2020). Urban auscultation; or, perceiving the action of the heart. *Places Journal*, April 2020. Accessed 13.5.2022. https://placesjournal.org/article/urban-auscultation-or-perceiving-the-action-of-the-heart/?cn-reloaded=1

Mattern, S. (2021). *A city is not a computer. Other urban intelligences*. Princeton: Princeton University Press.

McCann, E. J. (2007). Inequality and politics in the creative city-region: Questions of livability and state strategy. *International Journal of Urban and Regional Research, 31*(1), 188–196.

McNally, D., Price, L., & Crang, P. (Eds.). (2021). *Geographies of comfort*. London: Routledge. https://doi.org/10.4324/9781315557762

Merry, S. E. (1981). *Urban danger: Life in a neighborhood of strangers*, Philadelphia, PA: Temple University Press.

Middleton, J. (2011). Walking in the city: The geographies of everyday pedestrian practices. *Geography Compass, 5*(2), 90–105.

Middleton, J. (2021). *The walkable city. Dimensions of walking and overlapping walks of life*. New York: Routledge.

Miller, D. (1987). *Material culture and mass consumption*. Oxford: Blackwell.

Morris, B. (2004). What we talk about when we talk about 'walking in the city'. *Cultural Studies, 18*, 675–697.

Navaro-Yashin, Y. (2012). *The make-believe space. Affective geography in a post war polity*. Durham, NC: Duke University Press.

Niven, K. (2013). Affect. In M. D. Gellman, & J. R. Turner (Eds.), *Encyclopedia of behavioral medicine* (pp. 49–50). New York: Springer.

Norberg-Schulz, C. (1980). *Genius loci: Towards a phenomenology of architecture*. University of Minnesota: Academy Editions.

Nörenberg, H. (2018). Elementary affective sharing: The case of collective embarrassment. *Phänomenologische Forschungen, 1*, 129–150.

Nourani, F., Antonello, S. L., Govone, J. S., & Ceccato, V. (2020). Women and LGBTI youth as targets: Assessing transit safety in Rio Claro, Brazil. In V. Ceccato &

M. K. Nalla (Eds.), *Crime and fear in public places. Towards safe, inclusive and sustainable cities* (pp. 176–193). London: Routledge.

Nussbaum, M. C. (2013). *Political emotions*. Cambridge, MA: Harvard University Press.

Otter, C. (2008). *The Victorian eye: A political history of light and vision in Britain, 1800-1910*. Chicago, IL: University of Chicago Press.

Owen, J., Walker, A., & Ince, A. (2022). Editorial: Uncomfortable geographies. *Emotion, Space and Society*, *42*, 1–4.

Oznobikhina, I. I. (2021). What urban nightlife feels like: Atmospheric narratives and public spaces. In A. Elshater, H. Abusaada, & D. Rodwell (Eds.), *Transforming urban nightlife and the development of smart public spaces.* (pp. 40–53). Hershey, PA: IGI Global.

Pallasmaa, J. (2005). *The eyes of the skin*. Chichester: John Wiley & Sons.

Pallasmaa, J. (2014). Space, place and atmosphere. Emotion and peripherical perception in architectural experience. *Lebenswelt. Aesthetics and Philosophy of Experience*, *4*(1), 230–242. https://open-tdm.au.dk/blogs/materielkultur/wp-content/uploads/sites/11/2017/10/Pallasmaa_spaceplace-and-atmosphere.pdf

Payne-Frank, J. H. (2022). *The Aesthetics of Participation A Study of the Oslo Opera House*. Roskilde University. Roskilde. Unpublished PhD Thesis.

Pérez-Gómez, A. (2016). *Attunement: Architectural meaning after the crisis of modern science*. Cambridge, MA: MIT Press.

Peterson, M. (2021). *Atmospheric noise. The indefinite urbanism of Los Angeles*. Durham: Duke University Press.

Philippopoulos-Mihalopoulos, A. (2013). Atmospheres of law: Senses, affects, lawscapes. *Emotion, Space and Society*, *7*(1), 35–44.

Pickerill, J. (2015). Cold comfort? Reconceiving the practices of bathing in British self-build eco-homes. *Annals of the Association of American Geographers*, *105*(5), 1061–1077.

Pinder, D. (2005). *Visions of the city. Utopianism, power and politics in twentieth century urbanism*. London: Routledge.

Pinder, D. (2011). Errant paths: The poetics and politics of walking. *Environment and Planning D: Society and Space*, *29*(4), 672–692.

Powell, A. (2021). *Undoing optimization: Civic action in smart cities*. New Haven, CT: Yale University Press.

Price, L., McNally, D., & Crang, P. (2021). Towards geographies of comfort. In D. McNally, L. Price, & P. Crang (Eds.), *Geographies of comfort* (pp. 1–22). London: Routledge.

Relph, E. (1976). *Place and placelessness*. London: Pion.

Rendell, J., Penner, B., & Borden, I., (Eds.). (2000). *Gender space architecture: An interdisciplinary introduction*. London: Routledge.

Rhys-Taylor, A. (2014). Urban sensations: A retrospective of multisensory drift. In David Howes (Ed.), *A cultural history of the senses in the modern age* (pp. 55–76). London: Bloomsbury Publishing.

Riedel, F., & Torvinen, J. (Eds.). (2020). *Music as atmosphere: Collectice feelings and affective sounds*. New York: Routledge.

Rishbeth, C., & Rogaly, B. (2018). Sitting outside: Conviviality, self-care and the design of benches in urban public space. *Transactions of the Institute of British Geographers*, *43*, 284–298.

Rodaway, P. (1994). *Sensuous geographies*. London: Routledge.

Rokem, J., & Vaughan, L. (2018). Geographies of ethnic segregation in Stockholm: The role of mobility and co-presence in shaping the 'diverse' city. *Urban Studies*, 56(12), 2426–2446.

Rosa, H. (2019). *Resonance: A sociology of our relationship to the eorld*. Cambridge, MA: Polity Press.

Rosa, H. (2020). *The uncontrollability of the world*. Translated by James C. Wagner. Cambridge: Polity.

Rosenberger, R. (2020). On hostile design: Theoretical and empirical prospects. *Urban Studies*, 57(4), 883–893.

Rossi, A. (1982). *The architecture of the city* Cambridge, MA: MIT Press; first published in 1966.

Savic, S., & Savicic, G. (Eds.). (2012). *Unpleasant design*. Belgrade: G.L.O.R.I.A.

Schafer, R. M. (1985). Acoustic space. In D. Seamon, & R. Mugerauer (Eds.), *Dwelling, place, environment. Towards a phenomenology of person and world* (pp. 87–98). Dordrecht/Boston/Lancaster: Martinus Nijhoff Publishers.

Scherg, R. H. (2018). *Utryghed som fænomen. Er man tryg, hvis man ikke er utryg?* Copenhagen: Det Kriminalpræventive Råd.

Schivelbusch, W. (1987). The policing of street lighting. *Yale French Studies*, 72, 61–74.

Schivelbusch, W. (1988). *Disenchanted night: The industrialization of light in the nineteenth century*. Berkeley: The University of California Press.

Schönhammer, R. (2018). Atmosphere—the life of a place. The psychology of environment and design. In J. Weidinger (Ed.), *Designing atmosphere* (pp. 141–182). Berlin: Universitätsverlag der TU Berlin.

Schwabe, S. (2021). Order and atmospheric memory: Cleaning up the past, designing the future. *City and Society*, 33(3), 40–58.

Schwabe, S, Klaaborg, I. L., & Bille, M. submitted, Attuned visibility and the ambiguity of demanding public spaces in Copenhagen.

Seamon, D., & Mugerauer, R. (1985). *Dwelling, place and environment: Towards a phenomenology of person and world*. New York: Springer.

Seamon, D. (1979). *A geography of the lifeworld: Movement, rest, and encounter*. New York: St. Martin's.

Sendra, P., & Sennett, R. (2020). *Designing disorder: Experiments and disruptions in the city*. London: Verso.

Sennett, R. (2017). The open city. In Tigran Haas, & H. Westlund (Eds.), *The post-urban world. Emergent transformation of cities and regions in the innovative global economy* (pp. 97–105). London: Routledge.

Sennett, R. (2018). *Building and dwelling: Ethics for the city*. New York: Penguin.

Shaw, R. (2018). *The nocturnal city*. New York: Routledge.

Sheller, M., & Urry, J. (2006). The new mobilities paradigm. *Environment and Planning A: Economy and Space*, 38(2), 207–226.

Shibati, S. (2020). Transit safety among college students in Tokyo-Kanagawa, Japan: victimization, safety perceptions and preventive measures. In V. Ceccato & M. K. Nalla (Eds.), *Crime and fear in public places. Towards safe, inclusive and sustainable cities* (pp. 160–175). London: Routledge.

Simmel, G. (1971). *The stranger. Georg Simmel: On individuality and social forms* (pp. 143–150). Chicago: University of Chicago Press.

Simonsen, K., & Koefoed, L. M. (2020). *Geographies of embodiment: Critical phenomenology and the world of strangers* (p. 168). London: SAGE Publications.

Simpson, D. (2019). Liveability as metric regime. In K. L. Weiss (Ed.), *Critical city: The success and failure of the Danish welfare city* (pp. 27–49). Copenhagen: The Danish Architectural Press.

Simpson, D., Gimmel, K. Lonka, A. Jay, M., & Grootens, J. (2018). *Atlas of the Copenhagens*. Berlin: Ruby Press.

Simpson, P. (2017). A sense of the cycling environment: Felt experiences of infrastructure and atmospheres. *Environment and Planning A, 49*(2), 426–447.

Simpson, P. (2018). Elemental mobilities: Atmospheres, matter and cycling amid the weather-world. *Social and Cultural Geography, 9365,* 1–20.

Slaby, J. (2020). Atmospheres: Schmitz, Massumi and beyond. In F. Riedel & J. Torvinen (Eds.), *Music as Atmosphere. Collective Feelings and Affective Sounds* (pp. 274–285). London: Routledge.

Slaby, J., & von Scheve, C. (2019). Introduction. Affective societies: Key concepts. In J. Slaby & C. von Scheve (Eds.), *Affective societies: Key concepts* (pp. 1–24). Taylor and Francis Group: Routledge. https://doi.org/10.4324/9781351039260

Sloane, M., et al. (2016). *Tackling Social Inequalities in Public Lighting* (p. 41). LSE-based Configuring Light/Staging the Social research programme. London, UK: Configuring Light/Staging the Social.

Smith, K. (2012). From dividual and individual selves to porous subjects. *Australian Journal of Anthropology (The), 23*(1), 50–64.

Smith, M., Davidson, J., Cameron, L., & Bondi, L. (Eds.). (2009). *Emotion, place and culture*. Aldershot: Ashgate.

Solnit, R. (2001). *Wanderlust: A history of walking*. London: Granta.

Sørensen, T. F. (2015). More than a feeling: Towards an archaeology of atmosphere. *Emotion, Space and Society, 15,* 64–73. https://doi.org/10.1016/j.emospa.2013.12.009

Spadoni, R. (2020). What is film atmosphere? *Quarterly Review of Film and Video, 37*(1), 48–75.

Speck, J. (2013). *Walkable city: How downtown can save America, one step at a time*. New York: North Point Press.

Spinney, J. (2006). A place of sense: A kinaesthetic ethnography of cyclists on Mont Ventoux. *Environment and Planning D: Society and Space, 24,* 709–732.

Stavrides, S. (2018). Urban porosity and the right to a shared city. In S. Wolfrum, H. Stengel, F. Kurbasik, N. Kling, S. Dona, I. Mumm, & C. Zöhrer (Eds.), *Porous city* (pp. 32–37). Basel: Birkhaüser.

Stenslund, A. (2021). Collaging atmosphere: Exploring the architectural touch of the eye. *Environment and Planning B: Urban Analytics and City Science, 48*(9), 2761–2774.

Stenslund, A. (2023). *Atmosphere in urban design a workplace ethnography of an architecture practice*. London Routledge.

Stenslund, A., & Bille, M. (2021). Rendering atmosphere: Exploring the creative glue in an urban design studio. In M. Stender, C. Bech-Danielsen, & A. Landsverk Hagen (Eds.), *Architectural anthropology: Exploring lived space* (pp. 207–223). London: Routledge.

Stephens, A. C. (2015). Urban atmospheres: Feeling like a city? *International Political Sociology, 9,* 99–101. https://doi.org/10.1111/ips.12082

Stewart, K. (2011). Atmospheric attunements. *Environment and Planning D: Society and Space, 29*(3), 445–453.

Stoler, A. L. (Ed.). (2013). *Imperial debris: On ruins and ruination*. Durham: Duke University Press.

Strathern, M. (1988). *The gender of the gift: Problems with women and problems with society in Melanesia*. Berkeley, CA: University of California Press.

Sumartojo, S. (2022). How the city feels. In Sumartojo, S. (Ed.), *Lighting design in shared public spaces* (pp. 106–124). London: Routledge.

Sumartojo, S., & Pink, S. (2019). *Atmospheres and the experiential world.* New York: Routledge.

Taylor, C. (2007). *A secular age.* Cambridge: The Belknap Press of Harvard University Press.

Thelle, M., & Bille, M. (2020). Urban porosity and material contamination from cholera to COVID-19 in Copenhagen. *Journal for the History of Environment and Society, 5,* 171–180.

Thibaud, J.-P. (2011). The sensory fabric of urban ambiances. *Senses and Society, 6*(2), 203–215.

Thibaud, J.-P. (2013). Commented city walks. *Wi: Journal of Mobile Culture, 7*(1), 1–32.

Thibaud, J. P. (2015). The backstage of urban ambiances: When atmospheres pervade everyday experience. *Emotion, Space and Society, 15,* 39–46.

Till, K. E. (2012). Wounded cities: Memory-work and a place-based ethics of care. *Political Geography, 31,* 3–14.

Trigg, D. (2022). Atmospheres and shared emotions. In Dylan Trigg (Ed.), *Atmospheres and shared emotions* (pp. 1–14). New York: Routledge.

Tripathy, P., & McFarlane, C. (2022). Perceptions of atmosphere: Air, waste, and narratives of life and work in Mumbai. *Environment and Planning D: Society and Space, 40*(4), 664–682.

Tronto, J. C. (1993). *Moral boundaries: A political argument for an ethic of care.* New York: Routledge.

Tronto, J. C. (2015). *Who cares? How to reshape a democratic politics.* New York: Cornell University.

Tronto, J. C. (2019). Caring architecture. In Angelika Fitz, & Elke Krasny (Eds.), *Critical care: Architecture and urbanism for a broken planet* (pp. 26–32). Cambridge, MA: MIT Press.

Trudsø, A. L. (2013). *Koordinater.* Copenhagen: Rosinante.

Tuan, Y.-F. (1974). *Topophilia: A study of environmental perception, attitudes, and values.* Englewood Cliffs, NJ: Prentice-Hall Inc.

Urry, J. (2012). City life and the senses. In G. Bridge, & S. Watson (Eds.), *The new Blackwell companion to the city* (pp. 347–356). London: Blackwell.

Van Liempt, I., Van Aalst, I., & Schwanen, T. (2015). Introduction: Geographies of the urban night. *Urban Studies, 52*(3), 407–421.

Vannini, P. (Ed.). (2009). *The cultures of alternative mobilities. Routes less travelled.* Aldershot: Ashgate.

Vogels, I. (2008). Atmospheric metrics. A tool to quantify perceived atmosphere. In Westerink, J. H. D. M., Ouwerkerk, M., T. J. M. Overbeek, W. F. Pasveer, & de Ruyter, B. (Eds.), *Probing experience. From assessment of user emotions and behaviour to development of products* (Vol. 8, pp. 25–42). Dordrecht: Springer.

Wallace, D. F. (2009). *This is water. Some thoughts, delivered on a significant occasion, about living a compassionate life.* Boston, MA: Little, Brown & Company.

Walsh, C. (2008). The mosquito: A repellent response. *Youth Justice, 8,* 122–133.

Watson, A., Ward, J., & Fair, J. (2021). Staging atmosphere: Collective emotional labour on the film set. *Social & Cultural Geography, 22*(1), 76–96.

Welsh, Brandon C., Farrington, David P., & Douglas, S. (2021). *Effectiveness of street lighting in preventing crime in public places: An updated systematic review and meta-analysis.* Stockholm: Swedish National Council for Crime Prevention.

Westerink, J. H. D. M., Ouwerkerk, M., T. J. M. Overbeek, W. F. Pasveer, & de Ruyter, B. (Eds.), (2007). *Probing experience: From assessment of user emotions and behaviour to development of products*. Dordrecht: Springer.

Whyte, W. H. (1980). *The social life of small urban spaces*. New York: Project for small spaces.

Wigley, E., & Rose, G. (2020). Will the real smart city please make itself visible? In Katharine S. Willis, & Alessandro Aurigi (Eds.), *The Routledge companion to smart cities* (pp. 301–311). London: Routledge.

Wigley, M. (1998). The architecture of atmosphere. *Daidalos, 68*, 18–27.

Willis, K. S., & Aurigi, A. (Eds.). (2020). *The Routledge companion to smart cities* (1st ed.). New York. Routledge.

Wolf, B., & Julmi, C. (Eds.). (2020). *Die Macht der Atmosphären*. München: Verlag Karl Alber.

Wolfrum, S., Stengel, H., Kurbasik, F., Kling, N., Dona, S., Mumm, I., & Zöhrer, C. (Eds.). (2018). *Porous city*. Basel: Birkhaüser.

Yang, H., et al. (2022). Perceptions of safety in cities after dark. In S. Sumartojo (Eds.), *Lighting design in shared public spaces* (pp. 83–104). London: Routledge.

Zardini, M. (2005). *Sense of the city. An alternate approach to urbanism*. Montreal: Lars Müller Publishers.

Zeiderman, A., & Dawson, K. (2022). Urban futures: Idealization, capitalization, securitization. *City, 26*(2–3), 261–280.

Zerlang, M. (2008). Experiencing Israels Plads. *Nordic Journal of Architectural Research, 20*(1), 53–62.

Zukin, S. (2010). *Naked city. The death and life of authentic urban places*. Oxford: Oxford University Press.

Index

Page followed by n refer to notes.

15-Minute City 76, 84

Ahmed, S. 18–19, 44, 108
AI design 51–55, 121–122
Aker Brygge 40–42
Allen, J. 59–60
Ambient power 59–60, 64, 80, 120, 122
Amin, A. 5, 93
Anderson, B. 15
Anderson, S.L. 55
Atmosphere definition 14–16
Atmospheric memory 66
Atmospheric practices 57
Atmospherological power 8–9
Attunement 7–9, 11–12, 14–16, 27–28, 43, 48–49, 54, 70, 76, 89–90, 93, 109, 112
Attunement definition 16–20

Beaumont, M. 92
bell hooks 18, 81
Bellacasa, M.P. 100
Benjamin, W. 27
Berlant, L. 108
Bike culture 86–89
Bissell, D. 77
Black Square *see* Superkilen
Blågårds Plads 3–4, 120
Bollnow, O. 17
Borch, C. 46
Brunkebergstorg 57–60
Böhme, G. 15, 27, 56–57, 72n4

Camden bench 63–64
Care definition 96–99
CGI 14, 52–53, 55
Cole, T. 35
Collective effervescence 46
Covid-19 pandemic 5, 13, 25, 32, 82, 117

Danish Crime Prevention Council 108–109
Dawson, K. 122
Dissonance 18, 32, 40–44, 48
Duff, C. 84
Durkheim, E. 46

Estalella, A. 104
Experience economy viii, 6

Frederiksberg 111
Frogner 109

Gehl, J. 10–11
Genius loci 15, 72
Ghertner, A. 121

Hammershøi, V. 31
Heidegger, M. 16–17
Hommels, A. 72
Human-centric design 13, 93
Hygge 5

Ingold, T. 68, 70, 89
Intimate alienation 12
Israels Plads 37–40, 80–81

Jacobs, J. 28
Jensen, O.B. 82–83
Jiménez, A.C. 104

Kenopsia 26
Kierkegaard 89
Klingenberg, E. 97
Kungsholmen 110

Lacis, A. 27
Life-oriented city 7, 10–13, 34, 93, 114, 118

Index

Lonely Planet 3
Løgstrup, K.E. 100

Manning, E. 27
Massey, D. 27, 112
Mattern, S. 30
Mood *see also* attunement
Mosquito sound 122

New mobilities paradigm 76
Noise 28, 33, 34, 103, 106–107, 122
Nordhavnen 69
North West Park 56, 62–64, 100–103, 120
Northwest Copenhagen 87, 119
Nyboder 65
Nørrebro 3–4, 95, 106–108, 119

Oslo Central Station 51–52
Oslo Opera House 35, 78–80

Pallasmaa, J. 85
Pérez-Gómez, A. 81
Peripheral attachment 37, 39, 47
Peterson, M. 34
Pink, S. 16, 99
Porosity ix, 7–8, 13, 18, 21, 25, 27, 30–31, 42–44, 48, 57, 70, 75, 83, 93, 104, 108
Psychogeography 10, 12

Resonance 18, 28–31, 33, 45, 48, 104, 110, 113
Rhys-Taylor, A. 53
Rosa, H. 30, 104, 113

Safety definition 108–109

Scenographic city 52–53, 60, 62–64, 117, 120, 123
Seasonal Affective Disorder (SAD) 117, 122
Sergels Torg 90–91
SLA 55–56
Smart City *see* AI design
Solnit, R. 84, 89
Stenslund, A. 5–6, 55–56
Stephens, A.C. 88
Stockholm Royal Seaport 65–68
Stovner 61–62, 120
Sumartojo, S. 16, 98, 99
Superkilen 33
Surveillance 59–60, 91, 100–103, 120–121
Sørensen, T.F. 84

Taylor, C. 25
Thrift, N. 93
Throwntogetherness 112
Till, K. 113–114
Time Out 4
Tivoli 4
Trudsø, A.L. 65

Vesterbro 32–33, 42–43, 82

Walkable City 76, 84
Wallace, D.F. 71
Walsh, C. 122
Weather 68–70, 72
Wigley, M. 56

Zeiderman, A. 122
Zit bulb 122

Østerbro 95